Life in the Natural World

Investigating Life's Diversity

Third Edition

Steve L. O'Kane Jr.
Kimberly A. Cline-Brown

University of Northern Iowa

Cover image of starfish © 2012 Shutterstock, Inc.
Remaining cover images courtesy of Steve L. O'Kane Jr.

Kendall Hunt
publishing company

www.kendallhunt.com
Send all inquiries to:
4050 Westmark Drive
Dubuque, IA 52004-1840

ISBN 978-1-4652-0762-3

Printed in the United States of America
10 9 8 7 6 5 4 3 2

Contents

Preface v

Getting the Most Out of These Investigations vii

Lab 1: Your Body – A Habitat 1-1
- Get to Know Your Microscope 1-1
- Prepare a Wet Mount of Cheek Cells 1-3
- Observe Cheek Cells Using the Light Microscope 1-3
- They Have Us Outnumbered – Bacteria in Your Body 1-5
- Life on Your Face! 1-8

Lab 2: Water: Osmosis, Tonicity, and Transpiration 2-1
- Osmosis in Potato Cores 2-2
- Tonicity of *Elodea Cells* 2-5
- Transpiration From Plants 2-10

Lab 3: Life: Atmosphere and Energy 3-1
- Photosynthesis & Respiration 3-3
- The Color of Plants 3-4
- Storage Products of Photosynthesis 3-6
- Measure Dissolved Gasses 3-8

Lab 4: Estimating Biodiversity: The Species-Area Relationship 4-1

Lab 5: DNA Diversity and Its Measurement (two lab periods) 5-1
- Lab 1
  - DNA Isolation from Cheek Cells 5-3
  - PCR Amplification of the D1S80 VNTR Locus 5-5
- Lab 2
  - Gel Electrophoresis 5-6
  - So How Rare Is My Genotype? 5-7

Lab 6: Reproducer – The Process of Natural Selection 6-1
- Modeling Selection 6-2
- Changes in Genotypes and Allele Frequencies 6-8

Lab 7: Plant Reproduction – Getting Into the Future 7-1

Flower Morphology 7-4
Flower Power 7-7
Fruit and Seed Examination 7-10

Lab 8: Plant Nutrition and Symbiosis 8-1
Tomato Seedlings & Nutrients 8-2
Effect of Nitrogen Sources on Clover Plants 8-7
Examination of Root Nodules 8-8
Mycorrhizal Fungi 8-10

Lab 9: Soil – The World Beneath Your Feet (two lab periods)

Part I: Physical Properties 9-1
Texture 9-1
Field capacity 9-4
Wilting point 9-5
Available water 9-6
Other Physical Properties of Soil 9-7

Part II: Biological Properties – the Living Soil 9-11
Microbe Diversity 9-12
Estimating Microbial Population Numbers 9-13

Part III: Soil Animals 9-16
Soil Invertebrates 9-17
Tullgren-Burlese Funnels 9-18
Baermann Funnels 9-18
Group Comparisons 9-22
Illustrated Guide to Common Soil Animals 9-25

Lab 10: Aquatic Toxicology 10-1
Observations of *Daphnia* 10-2
The LD_{50} Test 10-2

Lab 11: Saving Biodiversity: An Example Using Threatened Communities 11-1

Lab 12: Puzzling the Past 12-1

Preface

Charles Darwin delighted in the diversity and complexity of the living world, extolling in his most famous work, *On the Origin of Species*, life's "endless forms most beautiful and most wonderful."

In this collection of laboratory investigations we have chosen to follow Darwin's lead by highlighting some of the many forms of life's biological diversity and complexity – and there are *many*! Evolution is, after all, an extremely creative process. You'll be examining diversity and complexity from the level of DNA, to that of species, and finally to how species are assembled into communities and how these communities are studied. You'll also be looking at some of the important mechanisms and natural systems that underlie, maintain, and change biological diversity. It is not much of an exaggeration to say that life as we know it, in all its myriad and beautiful forms, is about variation and interaction.

We, the authors of this laboratory manual, can claim only a limited degree of originality. We have read, tested, and culled numerous ideas and lab protocols to come up with those presented here. Much of what is presented here, however, was first dreamed up by other scientists and educators. (We hope that sometimes, at least, our original ideas do shine through.) These labs – really inquiries – are open to surprises. Your lab instructor will often not know the exact outcome of a particular laboratory experiment. In the real world, scientific experiments frequently don't work out or give the expected outcome. And that's as it should be! The fun part of science is the "eureka moment" when something new appears on the horizon.

These labs are designed to be stand-alone activities that are primarily self-directed. As such, they are not tied to a particular lecture course. You will find that your instructor will "teach" much less than you may be used to. Think of your instructor as a facilitator or as a resource for learning – the real learning is up to you. Said differently, the degree of learning you accomplish while conducting these investigations is almost entirely dependent on the effort you put into them, inside and outside the laboratory. Refer to the list of "how to" suggestions following this Preface for help on how to succeed and to get the most out of the course.

Ultimately, we hope that you'll find the biological world to be both intellectually intriguing and relevant to your life, no matter what your chosen path may be. If nothing else, our sincere hope is that you will end up being a better-informed citizen of the world which we share with the countless other beautiful and wonderful denizens of planet earth.

Acknowledgments

Several individuals were especially helpful to us in developing these investigations. First of all, Laura Walter was a significant contributor to the first and second editions of this manual, and her inspiration wafts through some of the current labs of this latest version. Our "Reproducer" lab was inspired by Sarah Hansvick's game "Survivor" found in the first edition of this manual. Mark Hafner (Louisiana State University) developed the lab on measuring biodiversity in threatened ecosystems, which is presented here with his kind permission. James Jurgenson provided expertise for the DNA labs. Dorothea Jurgenson and Michael Walter offered valuable suggestions on handling soil microorganisms. Virginia Berg greatly improved the lab activity related to photosynthesis and Maureen Clayton assisted with suggesting fieldwork methodology procedures for this lab. Kurt Pontasch and Dorothy Brecheisen developed the aquatic toxicology lab. Chris Schulte rendered most of the beautiful pen-and-ink drawings of organisms. The soil structure diagrams are courtesy of NASA. Further, we thank the many students who have helped us improve these investigations by their careful and thoughtful work with earlier versions of them.

Steve L. O'Kane Jr.
Kimberly A. Cline-Brown
May 2012

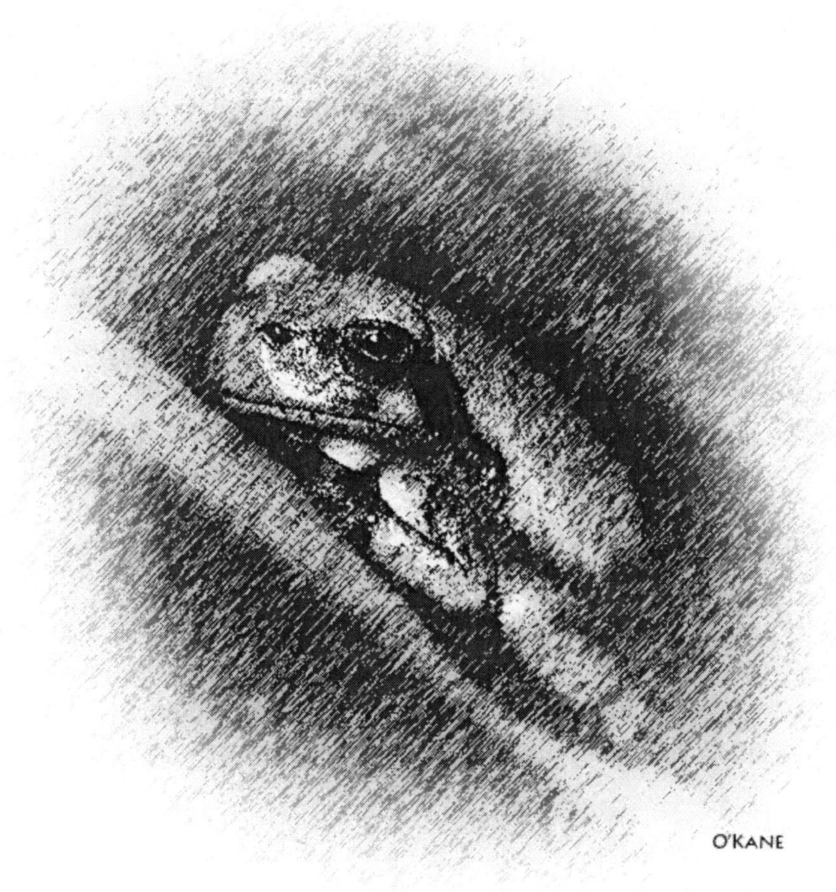

GETTING THE MOST OUT OF THESE INVESTIGATIONS

- Always (!) **read** the day's lab before you come to class – you may not understand all of it, but you will at least know what to expect. Reading the introduction to each lab is especially important so that you are familiar with the concepts you will be working with that day. (Check the syllabus to see what's on the schedule.)

- After conducting the lab, be sure you can **answer** the "Questions to Consider" that are sprinkled throughout. These questions are designed to help you think about the *meaning* of each lab. Thus, the answers may not necessarily be found in your readings and notes. Rather they require you to think about what you learned in lab class and apply this knowledge to develop an answer. The best time to work on these questions is shortly after the lab. At the minimum, try to have them done before your next lab period.

- **Ask** questions! Your instructor cannot read your mind and may be unaware that something is confusing. Try to stump your instructor with thoughtful questions. It makes the instructor's job much more interesting and helps to develop your reasoning skills – and it make's the instructor's day to have to say "I don't know!"

- **Notice** that introductions and explanatory comments are written in this font.

  Instructions are written in this font.

  Important vocabulary words are italicized.

- **Check off** the boxes (☐) provided in the procedures as you complete each step.

- **Work** together. Most of the activities in this manual are designed to be carried out by groups of two to four students. Get to know your lab partners – you may find that your individual strengths complement each other. But be sure that you carry your own weight!

- Most of all: **Have fun!** Science is exciting and enjoyable – no kidding.

LAB 1: YOUR BODY – A HABITAT

Throughout these investigations we will be exploring the diversity of living beings and the complexity of their internal and external interactions. We will, indeed, look in unexpected places. The first habitats we will explore are found in and on our own bodies.

You have heard it said that "no man is an island." From the point of view of many small organisms, this is not true. For them, every human *is* an island – one that provides the necessary habitat for them and for a host of other organisms, some beneficial, others detrimental, and some just along for the ride. A healthy person contains more cells of other species (mostly bacteria) than they do of human cells. About 90% of the cells in your body are not human. Wow. Most of your mass is you, not surprisingly, but the large majority of cells in your body are non-you. How could this be?

Think of yourself as a habitat – an island. What types of sustenance could other organisms obtain from your body? They could surreptitiously steal from your stomach or intestines some of the food you eat, quietly ingest your wastes or skin secretions, or even, on the sly, absorb nutrients from your bloodstream or organs. From their point of view, imagine how you would describe the types of habitat that exist in and on your body. Consider the availability of oxygen, of nutrients, acidity, humidity, temperature, light, physical space, and the activity of the body's defenses. Each species – no matter how simple or complex – has its own requirements. Variation in these conditions results in a diversity of habitats and, therefore, a diversity of organisms living on and in your own body – an island of exquisite biological diversity. How much more diverse must be the living beings in as simple a system as a small pond?

In today's investigation, you will explore your body as a natural habitat while gaining experience with one of the basic tools of biology – the microscope.

Get to Know Your Microscope

The modern light microscope is basically a pair of tubes with lenses at both ends. The lenses in the eyepieces are called *ocular lens*es and typically *magnify* an object ten times (10×). They make objects appear ten times larger than they actually are. The set of lenses nearest the microscope *stage* are the *objective lenses* (they are closest to the object under study). These lenses increase the magnification by amounts depending on which lens is used. The following table summarizes the magnification power of the usual combinations of ocular and objective lens combinations. (Your microscope may or may not have all of these.) Note that the **total magnification** of the microscope is found by multiplying the magnification of the ocular lens by that of the objective lens.

Objective Lens Type	Objective Lens Magnification	Total Magnification	Uses
Scanning	4× or 5×	40× or 50×	Locating specimens on the slide; preliminary focusing on small objects; and observing fine details of specimens that are visible to the naked eye.
Low power	10×	100×	Observing large cells, such as many plant cells, and initial focusing on smaller cells.
High-dry power	40×	400×	Observing smaller cells and smaller features within large cells.
Oil immersion	100×	1000×	Observing bacteria or very small structures within cells

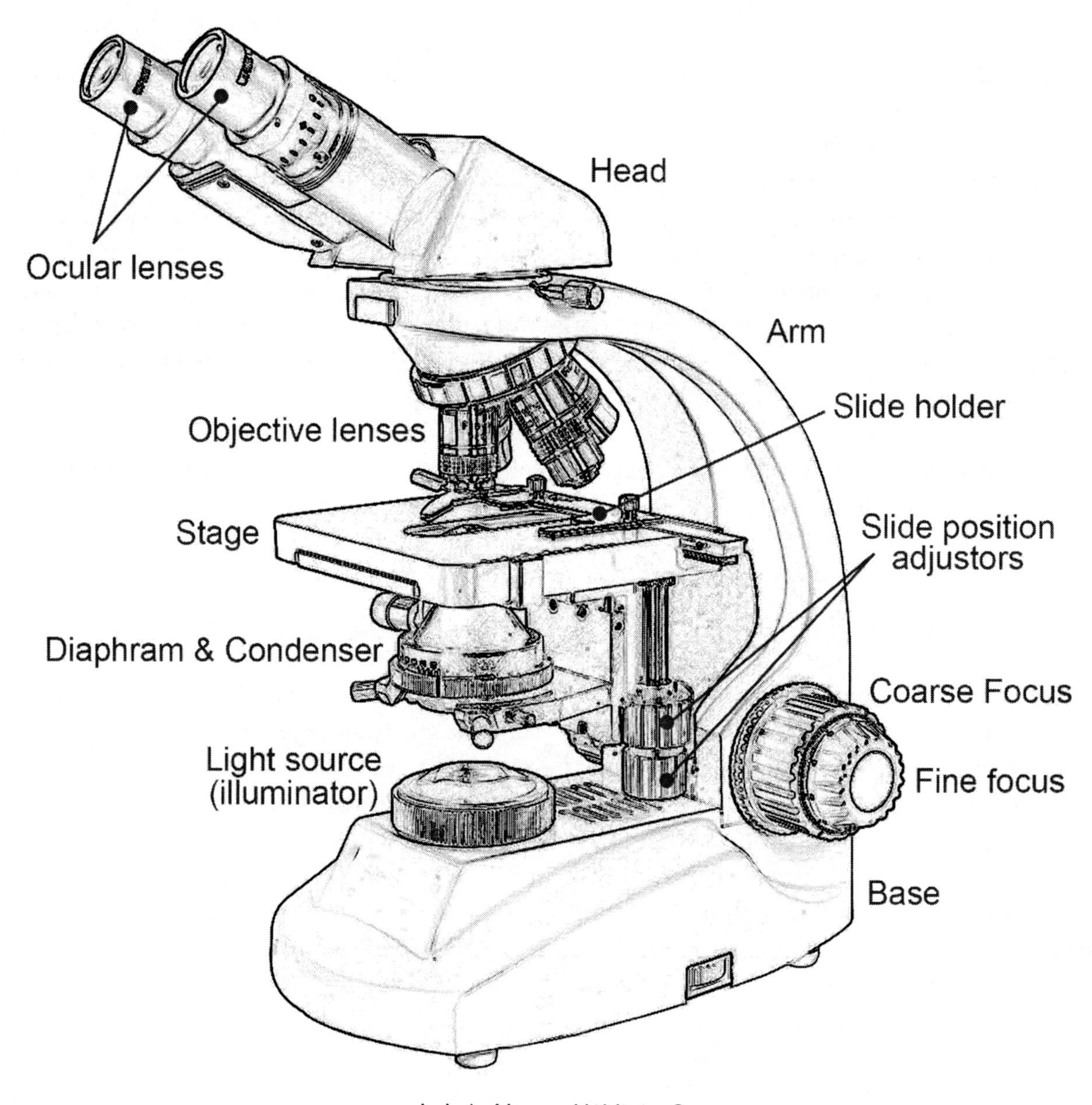

Activity 1: Prepare a wet mount of cheek cells

Procedure: (Check off each step as it is completed).

1. ☐ Obtain a clean glass slide and a plastic coverslip.

2. ☐ Place one small drop of methylene blue stain (from the dropper bottle at your bench) on the center of the slide.

3. ☐ Use a toothpick to vigorously scrape the inside of one group member's cheek. Stir the resulting "gunk" into the stain on your slide. Place the used toothpick in the waste container at your lab bench.

4. ☐ Touch one edge of the coverslip to the drop of stain, then slowly lower the coverslip until it is in place on the slide.

Activity 2: Observe cheek cells using the light microscope

Your instructor will "walk you through" the following procedures as a group. He or she will explain each of the terms as you follow the steps. Do not hesitate to ask questions!

Procedure:

1. ☐ Each pair or group of students should bring a microscope to their lab bench, holding the microscope upright, with one hand holding the arm and the other hand supporting the base. Place the microscope on the bench with the ocular lenses (eyepieces) facing you.

2. ☐ Raise the microscope stage using the large, coarse-focus knob.

3. ☐ Wipe <u>all</u> lenses with **lens paper**. Wipe the oil immersion lens **last**!

4. ☐ Place the objective lens on the lowest magnification. The magnification is marked on the side of the objective lens.

5. ☐ Lower the microscope stage a little using the large, coarse-focus knob.

6. ☐ Place your wet mount of cheek cells on the stage. Use the clamp(s) to hold the slide in place.

7. ☐ Move the slide until the specimen (the stained cheek gunk) is directly over the hole in the stage. Depending on your microscope, you do this using the set of knobs under the stage or by moving the slide or the stage by hand.

8. ☐ While looking from the side, use the coarse-focus knob to raise the stage until the cover slip is **almost** touching the objective.

9. ☐ Now look through the eyepieces and move the stage slowly away from you using the coarse focus knob, until the cheek cells come into focus. The cells will be darker blue than the background and look like a squashed balloon with a darker blob (the nucleus) inside.

10. ☐ Use the smaller, fine focus knob to sharpen the image. If other people look at the specimen, they should only use the fine focus knob to focus, **not** the course focus knob.

11. ☐ Now twist the low power (10×) objective into place above the specimen. Only use the fine focus knob to sharpen the image.

'Scope Note: When you begin to look at specimens, always start at a low magnification and then move to higher magnifications only after focusing and **centering** the specimen you are observing.

12. ☐ Adjust the light using the iris diaphragm lever located under the microscope stage. Less light often gives a more detailed image!

13. ☐ Move the slide or the mechanical stage to see different cells. Look for cells that are spread out so that you can see them individually. Notice which way the image moves as you move the specimen.

14. ☐ When you are focused on nice, well-spaced cells, slide the high-power (40×) objective into place **without changing the position of the stage** (that is, you should NOT have to use the course focus knob). Adjust the fine focus knob as necessary.

Cheek cells at 400x Magnification

Activity 3: They Have Us Outnumbered – Bacteria in Your Body

As has already been noted, there are more bacterial (and other) cells living on and inside your body than there are human cells in your body. Many are beneficial, helping us to digest our food for example, others are neither harmful nor helpful, while a few others may cause disease. Every human is, therefore, not just a single organism but a whole *community* of interacting organisms.

You can get a sense of just how many bacteria there are in your body by microscopically examining some of the plaque from between your teeth. Plaque bacteria obtain nutrients from the food you consume, and their acidic waste products can cause damage to your teeth. Dental plaque bacteria are also found in the arterial plaques of people with atherosclerosis and can cause dangerous infections in people with artificial heart valves or in the stints used to hold heart arteries open.

In non-agricultural and non-industrial societies people also have bacteria living on their teeth, but typically these do not cause tooth cavities (caries). It is only with the introduction of simple carbohydrates, like sugar, in our diets that dental caries appears. So, changing the kind of habitat your body provides via your diet can affect not just your health but also your bacterial passengers.

Procedure 1: Prepare a heat-fixed bacterial smear

1. ☐ Obtain a clean glass slide.

2. ☐ Take a sterile toothpick and scrape a small amount of material from the area between your teeth and thinly smear the “goo” on the slide on to a space about the size of a quarter. Place the used toothpick in the container provided.

3. ☐ Allow the smear to air-dry (about 1-2 minutes); then heat-fix it by picking up the slide with tongs and passing the slide (smear-side-up) five times through the hot (blue) portion of a Bunsen-burner flame.

4. ☐ Over the sink, apply enough methylene blue stain to cover the smear and leave it on for 30 seconds.

5. ☐ Pour off the stain and rinse <u>gently</u> under running water or slide washer

6. ☐ Dry the slide by blotting it, <u>without rubbing</u>, between two pieces of laboratory tissue (Kimwipes). Be sure the bottom of the slide (the side the smear is not on) is completely dry.

Procedure 2: Examine the stained smear under oil immersion

The oil in this technique acts like an extra lens, forming a clearer image by helping to keep the light from scattering. ***To avoid damaging the slide and/or the lens, use only the fine focus knob when working under oil immersion.***

1. ☐ Lower the stage using the coarse-focus knob and swing the objective lens to the lowest magnification.

2. ☐ Center the stained smear over the hole (the aperture) in the microscope stage.

'Scope Note: Don't center on the large chunky stuff, instead look for a lightly stained area. Thin tissue samples, especially when viewing single cell layers, yield better detail.

3. ☐ While looking from the side, raise the stage until the objective lens is **almost** touching the specimen.

4. ☐ While looking through the ocular lenses, lower the stage until the specimen comes into focus. Next move to the 10x lens and fine focus. Center your specimen under this lens.

5. ☐ Now swing the high-dry power objective (40×) into position, focus (with fine knob), and center your specimen. You may find some of your gum cells, which look just like cheek cells. If so, use them to get a good focus (Repeat steps 3 and 4 if necessary).

6. ☐ Adjust the light using the iris diaphragm. Have your Instructor take a look before proceeding to the next step.

7. ☐ Swing the high-dry objective aside <u>without changing the focus</u>.

8. ☐ Without lowering the stage, place a drop of immersion oil on the center of the smear. **STOP!** Make sure ONLY the oil immersion lens contacts the oil!

9. ☐ Twist the <u>oil immersion lens</u> into position. This objective should touch the oil.

10. ☐ Adjust the fine focus if necessary. Ask for help, if needed!

11. ☐ On the next page, sketch your results in the space provided. Try to illustrate the different shapes of bacteria that you found. Look carefully and closely!

12. ☐ <u>Clean up</u>: Place used slides in the container by the sink and clean the oil immersion lens with lens paper and lens cleaner.

13. ☐ Be sure to wipe up any oil on the stage of the microscope using lab tissues and ethanol.

Sketch of dental plaque bacteria

Questions to Consider

What types or shapes of cells did you observe? Was it easy or hard to distinguish between the types of cells?

How do the bacteria compare to your own (epithelial) cells that you saw in Activity 2 (wet mount of cheek cells)? Describe the size of your cells in relation to the bacteria? How many of your cells did you see relative to bacterial cells? You may want to draw this comparison below.

Activity 4: Life On Your Face!

Many people (as many as 100%, according to some sources) are hosts to microscopic mites (relatives of spiders) that live within the hair follicles and skin pores of our faces. These mites are highly specialized for living in skin pores: they have very reduced legs and thin, elongated bodies. In fact they can live nowhere else. The mites eat the contents of skin cells and the oils produced by sebaceous glands, but they do so little damage that most people do not even know that they harbor the mites.

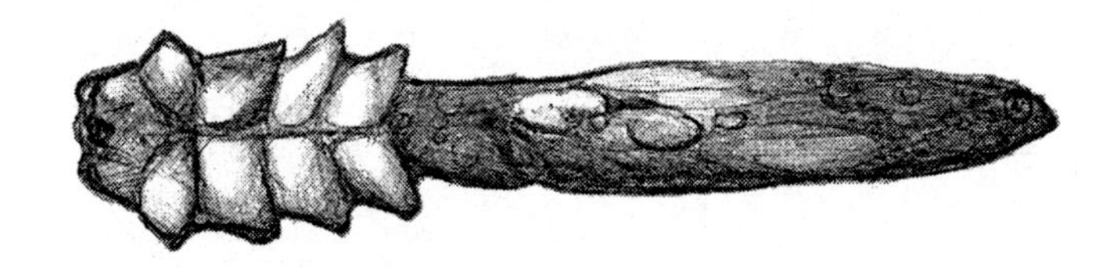

Underside of a facial mite

Procedure:

1. ☐ Place two small drops of immersion oil about 2 cm apart in the center of a clean glass slide.

2. ☐ Use a mirror to help with this step. Firmly squeeze the skin on the side of your nose until you see something ooze out. (Use a slow, firm pressing motion, like squeezing a pimple, because the mites are head-down deep in the pores and won't be found in the superficial skin oil.) Scrape off the "stuff" using a folded index card.

3. ☐ Use a toothpick to gently stir the stuff on the index card into one drop of oil on the slide. Add a cover slip.

4. ☐ With sterile forceps or your fingers, pluck a few eyebrows or eyelashes and add them to the other drop of oil, also covering them with a cover slip. Be sure you get a few "roots" (follicles), that's where the mites will be.

5. ☐ Search for the mites under the 10× low power objective (see steps 4–10 of Activity 1). Be sure to adjust the iris diaphragm to improve contrast. Look for moving legs! Once you have a mite in good focus, switch to the high-dry power (40×) objective for a closer look.

6. ☐ If you find mites, let your instructor know. He or she may wish to share your results with others. (Don't be shy – mites are *not* a sign of dirtiness.)

7. ☐ If you can't find any mites, don't give up until your lab partner and the instructor have looked at your slide. Don't be discouraged if you don't find any mites. Not everyone has them and they *are* hard to find.

8. ☐ Your instructor will count the number of students who found mites. Jot down the class data in the table below.

# of students sampled	# with mites

Questions to Consider

What percentage of students harbored mites?

Some sources report that men are more likely to harbor mites than women. Suggest possible reasons for this.

Until recently, people probably had more mites than they currently do. What are possible reasons for this?

What characteristics of facial pore mites allow them to successfully inhabit facial pores?

Name: ______________________________ Instructor's Name: _____________

Normal Lab Day and Time: _______________

Natural World Lab Response Sheet
Lab 1: Your Body A Habitat

After completing your lab , please answer the questions below. Then tear out this page and hand it in to your Instructor.

1. Summarize the main concepts behind today's' lab.

2. If you are using a 15X objective lens and the same ocular lens we have on our class microscopes, what is the total magnification? Be sure to show your work (write out the mathematical equation).

3. What characteristics make the human mouth a good environment for certain bacteria to live?

4. Rank today's lab from 0 (poor) to 10 (excellent). WHY did you choose this rating?

LAB 2: WATER: OSMOSIS, TONICITY, AND TRANSPIRATION

Water moves into and out of organisms. Your body is about 65% water: you drink, you sweat, you urinate, and yet your body continues to contain the same amount of water. In plants, water enters through the roots, moves through the plant's conducting tissue, and ultimately evaporates from the surfaces of leaves. Other than through evaporation and normal elimination (sweat and urine for example), water doesn't normally "leak" out of an organism's cells. Why not? And how does water get into an organism's cells? Put simply, water has to move through the *cell membrane*, that surrounds all cells, to get into and out of a cell. The process of water movement across a membrane is called *osmosis*.

But how does water "know" whether to move into or out of the cell? It turns out that water moves from areas of high concentration to areas of low concentration. (This assumes that neither the cells nor the water are under pressure.) Physiologists would say that water moves from areas of high *water potential* to areas of lower water potential. Here's a thought experiment: (1) take two beakers of water each containing 500 ml (milliliters) of water. (2) to one of the beakers add 50 g (grams) of a *solute* (disolved substance) such as salt. After the salt dissolves, the beaker will contain slightly more than 500 ml of liquid. (3) Remove exactly 1 ml of liquid from each beaker. Which milliliter of liquid contains more water? You should recognize that the milliliter of pure water has more water than does the milliliter of salt water. In other words, water is more concentrated when there are no solutes in it. Pure water has a *higher* water potential than does salt water.

Once we understand how molecules move along concentration gradients, we can apply this knowledge to all sorts of situations. One of these is *tonicity*, the fact that a cell's shape and "firmness" changes in response to its internal pressure as it gains or loses water depending on the type of solution it finds itself. Imagine what would happen to a cell if it was placed in a solution that had a higher solute concentration (and thus lower water concentration) than it did? Knowing what you know about water potentials, would water move from the solution into the cell, or out of the cell and into the solution? When a solution has a high solute concentration, the solution is said to be *hypertonic* (*hyper*, above + *ton*, tension) to the cell. In *hypotonic* (*hypo*, below) solutions, which have less solute concentration (and thus more water or solvent), water will move into the cell – causing it to become "bloated." When the amount of solute in the cell and solution are the same, the solution is *isotonic* (*iso*, same) to the cell and there is no change in the volume of water in the cell, so the cell's shape stays the same.

Let's consider cell membranes. Membranes are *semipermeable*. This means that some dissolved substances (solutes) move across the membrane freely, some are slowed, and some cannot cross the membrane at all. Water can move across membranes very freely. Salts, sugars, and most other dissolved substances move across membranes much more slowly or not at all. Cells contain salts, sugars, and other solutes. Perhaps you've tasted your own blood. Tasted salty didn't it? What would happen if you were to place one of your blood cells in a beaker containing pure water? Will water move into or out of the cell? In the first part of this lab, you will examine the movement of water across cell membranes as it is affected by solute concentration. You will use different concentrations of mannitol, a sugar alcohol that does not freely move across cell membranes, to examine the water potential of potato cells.

Experiment 1 – Osmosis in Potato Cores

Procedure: Work in pairs of students.

1. ☐ Label four test tubes with tape for *each* of the following mannitol concentrations 0%, 3.5%,7%, and 10.5%.

2. ☐ Wet a paper towel with distilled water to place potato pieces on.

3. ☐ Make eight full-length cores of a potato using a #1 cork borer. Cut *both* ends of the cores flat with a razor blade (don't include any "skin"). Do not touch the cores with your hands – use tweezers to manipulate the potato cores. Cut the cores so that they are all *about* the same length. (While you're at it, make a mental note of the firmness of the cores.)

4. ☐ Carefully measure and record the length of *each* of your cores. Be sure to measure to the nearest *half* mm:

Mannitol Concentration

	0%	**3.5%**	**7%**	**10.5%**
Core 1	____mm	____mm	____mm	____mm
Core 2	____mm	____mm	____mm	____mm
Total	____mm	____mm	____mm	____mm

5. ☐ Place two potato cores into each test tube.

6. ☐ Pour the appropriate mannitol solution into the tube until it is about 1 cm above the potato core.

7. ☐ Check the time and write it here ________. Wait 45 minutes.

While you are waiting, move onto the next major part of the lab called Experiment 2. Come back to this point after 45 minutes. Don't forget!

8. ☐ After 45 minutes, take the cores out of the test tubes and dry them gently with a paper towel. Measure the length of each core carefully to the nearest half mm. Record lengths below. Also record the amount of change in total length in mm. *Be sure to indicate whether the change is positive or negative*. For example, 2.5 or -1.0 or -0.35.

Mannitol Concentration

	0%	**3.5%**	**7%**	**10.5%**
Core 1	____mm	____mm	____mm	____mm
Core 2	____mm	____mm	____mm	____mm
Total	____mm	____mm	____mm	____mm
Change	____mm	____mm	____mm	____mm

Did you notice any changes in the firmness of any of the cores? Record if the cores were the same, firmer, or less firm here.

0%____________ 3.5%____________ 7%____________ 10.5%____________

9. ☐ Calculate the average percent change in the cores' lengths. Do this by dividing the *change* by the *original* total length and multiplying the result by 100. Be sure to keep track if the change is negative or positive. Record your results below.

Mannitol Concentration

	0%	**3.5%**	**7%**	**10.5%**
% Change	____%	____%	____%	____%

10. ☐ Plot your results on the graph below. Connect the plotted points with a best-fit straight line.

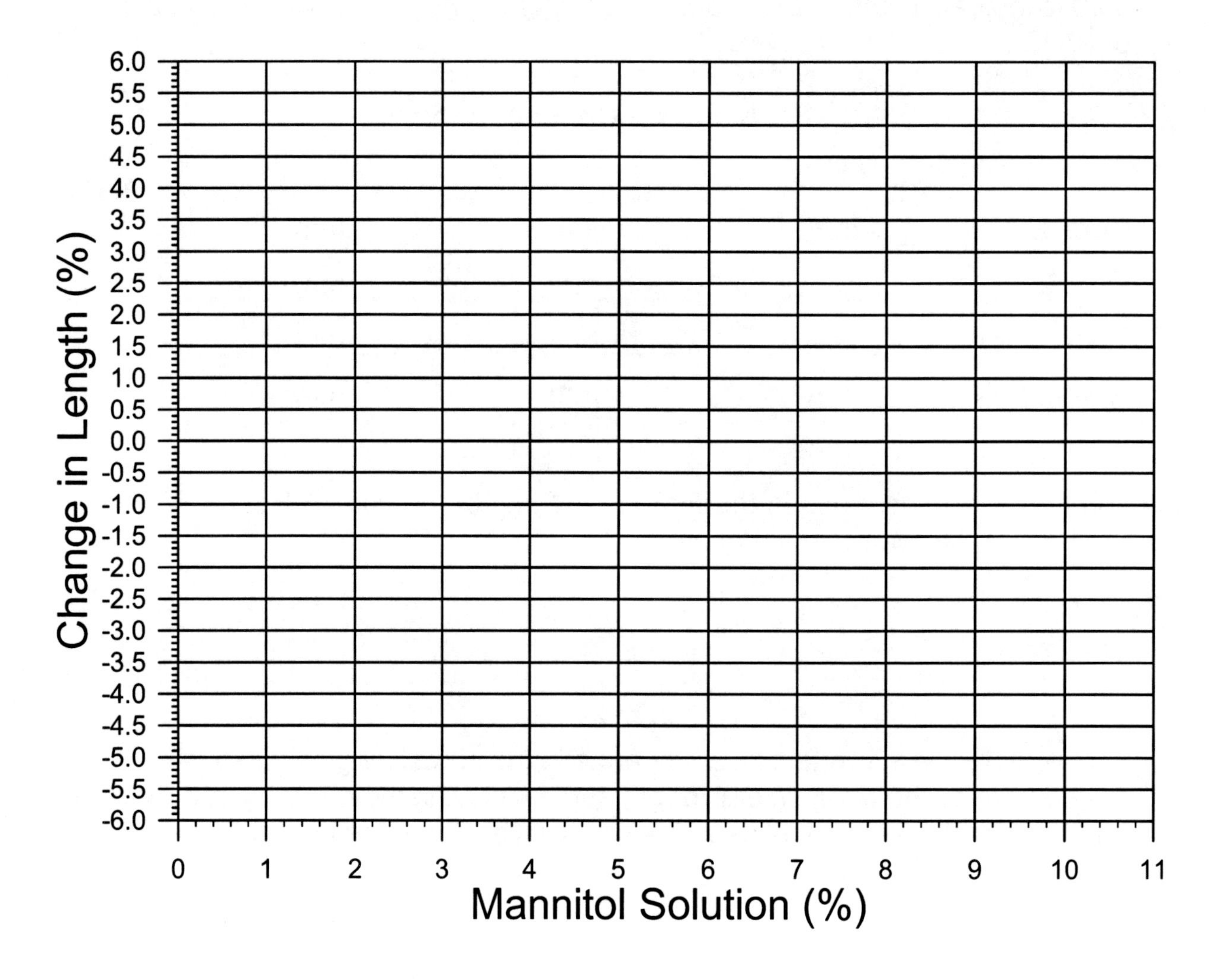

11. ☐ So what, why did we do this?

The "so what" of this experiment is that it illustrates a way to determine the water potential of potato cells. Look at the point on your graph where the line crosses zero on the Change in Length Axis. That's the point where there is no *net* movement of water into and out of the potato cells – the potato cores won't change in length at this mannitol concentration. In other words, the water potential is the same inside and outside the cell at this mannitol concentration.

Based on your results, what concentration of mannitol equals the water potential of potato cells? Write your answer here.

__________ % Mannitol   This is the percent mannitol that equals the water potential of the potato cells.

Questions to Consider

Why did you measure two, rather than just one potato core for each mannitol concentration?

Why does the potato swell when placed in a solution with no mannitol?

Sea water has about the same amount of salt in it as does your body. You've noticed that if you stay in the bath tub or swimming pool for a long time your finger tips and palms get wrinkly. In which kind of water will your skin become more wrinkly, sea water or bath water? Why?

If you water a plant with very salty water, will water move into or out of the roots?

Some plants and animals can live in very salty conditions. Do you suppose their cells have more salts than do organisms living in less salty environments? Justify your answer.

Experiment 2 - Tonicity of *Elodea* cells

In this part of the lab, you will examine, directly, the effects of salt concentration on osmosis in individual cells. In plants, most of the cell's volume is taken up by the vacuole, a storage organelle. Plant cells, unlike animal cells, have rigid cell walls. (See the Figure below.) Just inside the cell wall is the cell membrane. The whole thing is rather like a water balloon stuffed inside a box, except that water and most solutes can easily pass through the box. If water moves into the cell, the cell expands and exerts pressure on the surrounding cell wall. Without the cell wall, the cell would actually burst! In fact, many animal cells will burst if placed in pure water. If placed in salty water, water moves out of the cell. The cell contracts and pulls away from the cell wall. Eventually the cell will collapse, a process called *plasmolysis*, if the water is salty enough. You'll be using the water plant *Elodea* to illustrate these principles. The Figure below illustrates one *Elodea* cell.

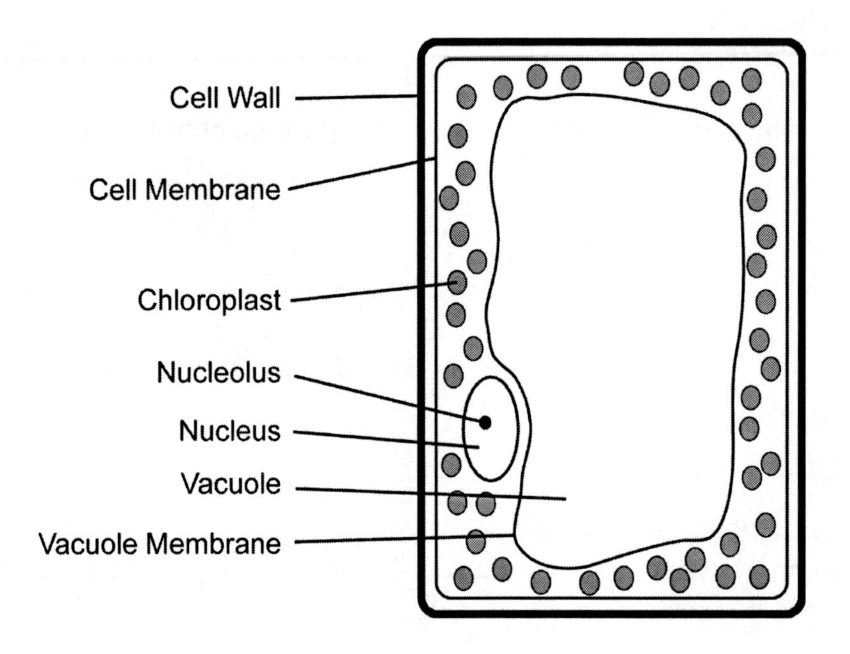

Procedure: Work in pairs of students.

1. ☐ Make a wet mount of an *Elodea* (a water plant) leaf.

2. ☐ Observe a thin area of the leaf under a microscope. Start at low power and move to higher power. Be sure to use high enough power so that you can see individual cells clearly. In the space below, sketch a few cells as seen at 100× or 400× magnification. Make your drawing a reasonable size and label the parts of the cell that you can see. The small green organelles you see surrounding the vacuole are chloroplasts.

   'Scope Note: The nucleus and nucleolus are not usually seen without special staining. The vacuole membrane may also not be visible. Be sure to note where these *should* be and draw them on your diagram. If you are unsure of where to draw them, please ask your instructor, or look at the diagram above to help..

3. ☐ Now, add four drops of 5% salt solution to one edge of the cover slip. Draw the solution under the cover slip by placing a flat edge of paper towel on the opposite side of the cover slip. This will replace the water with salt water.

4. ☐ *Immediately* begin observing the cells using the microscope. Observe them for a while so that you can see changes as they occur. Provide a drawing of a few cells. See note above in Step 2.

What happened to the plant cells? Was water moving into or out of the cell? Describe why/how water moved into or out of the cell.

In your group, discuss what you think would happen if you were to now place the *Elodea* cell in 100% (distilled) water. How might the cell now look? Write your answer here.

5. ☐ From the choices on the instructor's table, replace the salt solution with a solution that you think will rehydrate the *Elodea* cells by adding several drops of this solution to one side of the cover slip. Draw it through. Repeat at least once

more. Immediately begin observing the cells using the microscope. Observe them for a while so that you can see changes as they occur.

Solution Chosen: ______________________________

What happened to the cells? Why did this happen?

As a group, answer the questions below about your solution's effect on the cells. Choose **two** people from your group who will ***briefly*** tell the class about the effects of your group's solution choice on the cell.

1) How did your *Elodea* cells look after your solution was added? Draw and label a diagram.

2) Did your solution choice hydrate your cell? Why or why not?

Questions to Consider

What type of solution could hydrate an *Elodea* cell after it had been exposed to salt water?

What would happen to a fresh water fish if you put it into salt water?

What tonicity is the salt solution to the freshwater fish?

Conversely, what would happen to a salt water fish if you put it into fresh water?

What tonicity is the freshwater to the salt water fish?

Why does it hurt if you splash pure water in your eyes?

In a drought, do plants suffer more in salty soil or in soil low in salts? Why?

Experiment 3: Transpiration From Plants

Loss of water as a vapor from the surfaces of plants is called *transpiration*. Most of the water lost from plants is lost from the leaf *epidermis* through special pores called *stomates*. (See the Figure below.) The stomate itself is bordered by two *guard cells* that change shape in response to changes in available water and light (they typically close at night). When guard cells are full of water they are under pressure and move apart from one another in the center, forming the stomatal opening.

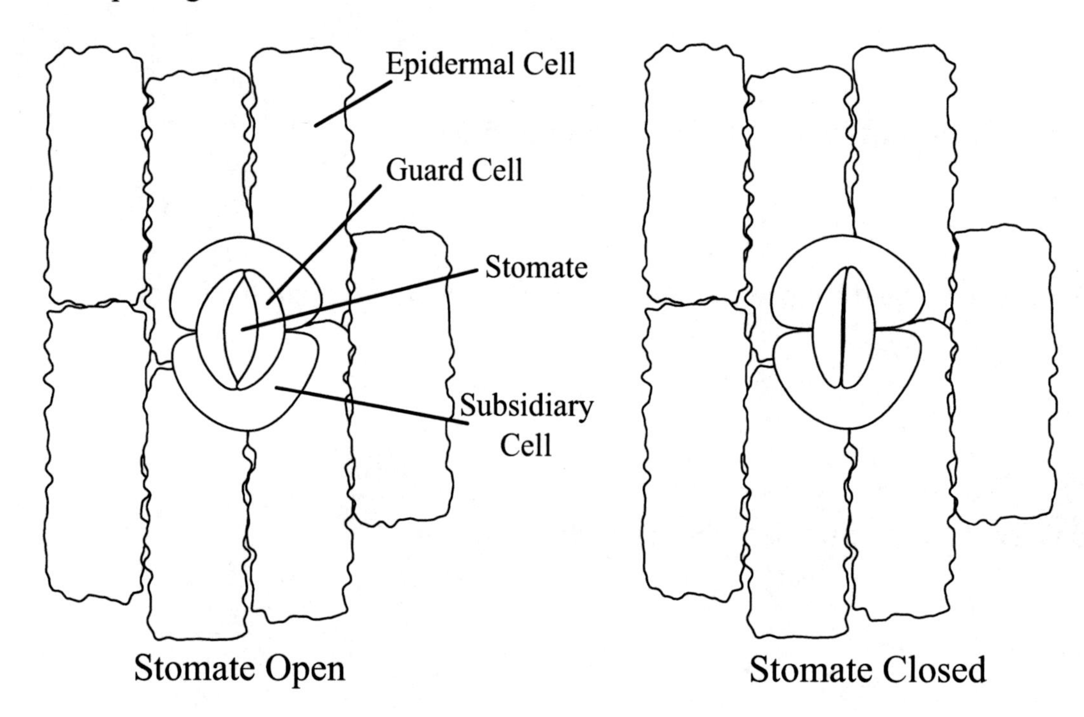

Plants are an integral component of earth's water balance. They literally "connect" the soil to the atmosphere. Water in the soil moves into a plant's roots, moves through the plant's water conducting tissue, the *xylem*, and exits the plant as vapor from the leaves. Evaporated water eventually falls back to earth as precipitation and re-enters the soil. Consider this: given that all nutrients are dissolved in water, how do the nutrients get to the top of the plant and to the plant's leaves? Evaporation of water from a plant's leaves "pulls" water up from the roots. This water contains the dissolved nutrients. If water didn't evaporate from the leaves of the plant, nutrients could not move through the plant! In this portion of the lab, you will examine leaf epidermis and the stomates embedded in it.

Procedure: Work in pairs.

1. ☐ Remove two healthy looking leaves from one of the plant types.

2. ☐ Paint thin strips of clear nail polish beside each other on the *underside* of one of the leaf pieces. Do the same to the *top side* of the other leaf piece.

3. ☐ Let the polish dry. Peel the polish off of the leaves. You might have to use forceps to do this.

   You've now made *leaf impressions* that are actually molds of the different kinds of cells on the leaf surfaces.

4. ☐ Label two microscope slides: one for the underside, one for the top.

5. ☐ Break off a small piece of each leaf impression and place it *leaf-side-up* on the appropriately labeled microscope slide. Try to use a piece that is as flat as possible so it doesn't brush against the objective lens of the microscope. If your imprint is curling on the edges or not lying flat, place a coverslip on top of the leaf impression to flatten it.

6. ☐ Starting at low power, examine each sample for stomates. Examine the underside first. Once you have located some stomates, move to higher power until you have a good view of one or a few stomates. Be sure to focus up and down to get a feel for the three-dimensional nature of the cells.

   Record how many stomates you can count in one field of view at 400× magnification:

   underside: ______________________ topside:_________________________

7. ☐ Draw a stomate along with the surrounding cells in the space below. Be sure to label the parts of your drawing.

Questions to Consider

What function do stomates play in regulating water loss in a plant?

Why would plants “want” to evaporate water from their leaf surfaces?

Plants use carbon dioxide (CO_2) in photosynthesis. CO_2 enters leaves through the stomates. What will happen to a plant’s ability to photosynthesize if it is stressed for water? Explain.

Would you expect to find stomates on an *Elodea* leaf? (Remember: it is a submerged water plant.) How does *Elodea* obtain carbon dioxide for photosynthesis?

On which side of a leaf did you find the most stomates? Think of a reason why this is true. Hint: which side of the leaf will be warmer, perhaps even hot, during the day?

Where would you look for stomates on a waterlily leaf?

Name: ______________________________ Instructor's Name: _____________

Normal Lab Day and Time: _______________

Natural World Lab Response Sheet

Lab 2: Osmosis, Tonicity & Transpiration

1. What was the main idea behind today's lab?

2. Use the diagram to help you answer the questions below.

You place several red blood cells (RBCs) in a beaker filled with a solution. What direction will water flow? What is the tonicity of the blood cells to the solution in the beaker? Answer these questions by circling the correct answer in the statements below.

Solution In Beaker:
95% H20
5% solute

Red Blood Cells:
85% H2O
15% Solute

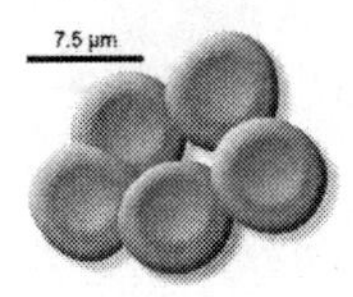

Water flows … (circle one) ***into the cell*** ***out of the cell*** ***remains the same***

The RBC is (circle one) ***isotonic*** ***hypotonic*** ***hypertonic*** to the beaker solution.

3. Using the membrane transportation concepts (e.g., osmosis) from today's lab, should you drink kool-aid when you are thirsty? Does it matter if it is sugar-free or not? Remember to *thoroughly* explain why or why not.

4. Rank today's lab from 0 (poor) to 10 (excellent). *Why?*

LAB 3: LIFE: ATMOSPHERE AND ENERGY

Back in the nineteen sixties, a chemist named James Lovelock, while looking for signs of life on Mars by analyzing its atmosphere, came to a startling conjecture about Earth. His idea, called the Gaia hypothesis, "suggests that the entire range of living matter on Earth, from whales to viruses, can be regarded as a single entity, capable of manipulating its environment to suit its needs."

Environmentally minded people tended to embrace the Gaia hypothesis with almost spiritual fervor. The majority of *ecologists* (scientists who study the interrelationships among organisms and their environments), however, did not think that the earth is actually a single organism or "superorganism." These ecologists, however, did and do find value in Lovelock's idea and now see it as an interesting metaphor that is not literally true. Our planet does indeed function, in many ways, like a giant superorganism, but one which cannot actually "know" its own needs. For one thing Earth "breathes" in the sense that organisms on Earth continually recreate the atmosphere that supports life – breathing in carbon dioxide and expelling oxygen.

Plants, cyanobacteria ("blue-green algae"), and photosynthetic algae in the process of *photosynthesis* use energy harvested from the sun to form high-energy carbon compounds that store captured (*fixed*) sunlight. Plants build these compounds by *fixing* carbon dioxide (CO_2) from the atmosphere and by splitting water molecules obtained from the soil. The sun's energy is then stored in the bonds that hold the carbon molecules together in these compounds, as in the first main product of photosynthesis, the sugar we call glucose ($C_6H_{12}O_6$). Plants then use this energy to build stems, leaves, roots, flowers, and fruits; to generate attractive colors and potent toxins; and to emerge as tiny shoots from buried seeds. In the process of light-harvesting, a molecule of water (H_2O) is split and oxygen (O_2) gas is given off. All of earth's breathable oxygen comes from water that has been split by plants and other photosynthetic organisms!

Photosynthesis radically changed the Earth's atmosphere. Before the cyanobacteria , then the algae, and lastly the land plants came into being, earth's atmosphere was predominantly composed of nitrogen and a large component of carbon dioxide. Now, while nitrogen still accounts for about 79% of the atmosphere, oxygen makes up most of the difference (~21%). CO_2 makes up only a little more than 0.03%.

This is fortuitous for all of us oxygen breathers. We harvest the energy plants store in carbon compounds and subject it to the "slow burn" called *respiration*, which *requires oxygen*. So, we end up with the energy we need to find *more* plants to eat, build bodies, find mates, escape from danger, keep warm, and manufacture the neurotransmitters it takes to compose or read this paragraph. In the process of respiration, complex carbon molecules are broken down into carbon dioxide, which is released into the atmosphere and can also be dissolved in water. <u>All</u> organisms respire, including plants. In fact, <u>most</u> of the energy fixed by plants is lost through their own respiration. **But the *net* effect in plants is: more oxygen is produced than is used and more CO_2 consumed than released.** Consider the illustration below where the processes of photosynthesis and respiration are summarized.

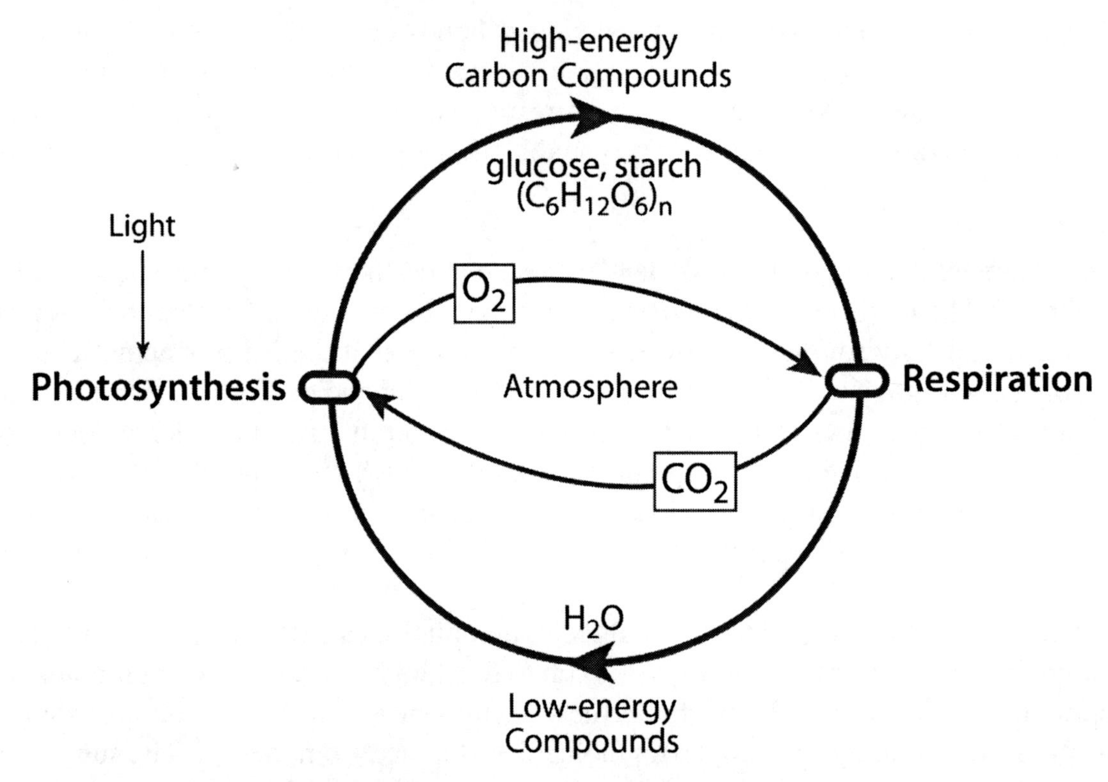

Notice that photosynthesis builds *high energy compounds from low energy compounds*. Conversely, respiration breaks high energy compounds down into low energy compounds. Photosynthesis uses CO_2 and releases O_2. Respiration uses O_2 and releases CO_2. In many ways, these two processes are the reverse of one another! This is also seen when you write out the equations for each process.

Photosynthesis: $6CO_2 + 6H_2O + \text{light} \rightarrow C_6H_{12}O_6 + 6O_2$

Respiration: $C_6H_{12}O_6 + 6O_2 \rightarrow 6CO_2 + 6H_2O + \text{energy}$

The bottom line is that living beings, the sun's energy, and the atmosphere are linked by the processes of photosynthesis and respiration.

In Activity 1of this lab, you will set up an experiment to investigate what is produced and what is consumed in both respiration and photosynthesis by experimenting with aquatic plants and fish. Activity 2 will examine the use of light by plants. Activity 3 will examine the formation of starch (a compound that stores carbon) in plant leaves. And, Activity 4 will return to the experiment set up in Activity 1 to measure your results.

Activity 1: Photosynthesis & Respiration (Setup)

Procedure – Setting up the Experiment: (Modeled on a protocol in Enger, E. E. and R. Otto. 1996. Laboratory Manual. Concepts in Biology, 8th ed. Wm. C. Brown Pubs., Dubuque, Iowa.)

1. ☐ Work in groups of 3-4 students.

2. ☐ Fill four 100 ml beakers up to the 80 ml line with **aged** tap water. Pour the water down the side of the beaker to avoid getting air bubbles in it. Put your group name or initials on each of the beakers. Aging allows the chlorine dissolved in tap water to diffuse out of the water.

3. ☐ Leave one beaker of water as it is. Label this beaker "Control."

4. ☐ Pack *each* of two of the 100 ml beakers with a 15 cm sprig of fresh-looking water plant (*Elodea*). Label one beaker "Dark" and the other "Light."

5. ☐ Net a small goldfish and gently place it in the fourth beaker. Label the beaker "Fish."

6. ☐ Cover each beaker with a piece of plastic wrap and wrap a rubber band around the flask to hold it in place.

7. ☐ Place the beaker labeled "Dark" in the cupboard under your lab bench. Put the other three beakers under fluorescent grow lights.

8. ☐ After sixty minutes, you will test the water from each of the beakers for its dissolved CO_2 and O_2 content. Note the time, and move on to the next section.

9. ☐ Using your knowledge of photosynthesis and respiration, make some predictions about your results.

   Which of the beakers will have

   -the highest dissolved CO_2 content? ____________________
   -the lowest dissolved CO_2 content? ____________________
   -the highest dissolved O_2 content? ____________________
   -the lowest dissolved O_2 content? ____________________

   Which will have a higher CO_2 content, the Control or the Dark Plant? Why?

Activity 2: The Color of Plants

What color are plants? Green of course! But *why* are they green? That's the real question and the answer to that question is important. You know by now that plants absorb sunlight and convert *some* of the energy of sunlight into the high energy chemical bonds of carbon compounds like glucose (a sugar) and starch. Not all wavelengths (colors) of light are equally useful for photosynthesis. You're about to discover which wavelengths are most useful and which are not.

Procedure: Wavelengths of Light Used in Photosynthesis

1. ☐ Work in groups of 3-4 students. **Warning!** The green plant extract *will* stain your clothes.

2. ☐ Obtain a deep glass petri dish containing about 50 ml of plant extract.

   Your instructor used a blender to liquify leaves in 80% ethanol. Chlorophyll easily dissolves in ethanol. The liquid was then filtered to produce an extract high in *chlorophyll*, the green plant pigment that captures sunlight.

3. ☐ Hold a "prism slide" about an inch from your eye and look at a microscope illuminator (or bulb masked with a slit). You should see a bright "rainbow" off to one side. Adjust the intensity of the light so that each color is clearly visible with no white washed-out areas between the colors.

4. ☐ Take careful note of the colors of this rainbow. Notice that the white light of the illuminator is broken into its component colors of red, orange, yellow, green, blue, indigo, and violet (and everything in between). Notice, too, the intensity of each color (wavelength) of light.

5. ☐ Now, hold your petri dish between the prism and the illuminator. You will see that certain colors disappear altogether from the "rainbow," and others will lessen in intensity.

6. ☐ Start at one end of the color spectrum and *carefully* note which colors (1) disappear or lessen in intensity when they are "filtered" by chlorophyll and (2) which colors are unaffected (you can still see -- so used LESS in photosynthesis). In the table below, record your observations.

Red ______________________ Green ______________________

Orange ______________________ Blue ______________________

Yellow ______________________ Violet ______________________

Colors which are absorbed by chlorophyll provide the energy to form high energy bonds in photosynthesis. Said another way, chlorophyll absorbs particular wavelengths of light and passes the energy of this light off to other compounds, which use the energy to fix carbon dioxide. There is a way to directly observe chlorophyll handing off energy. Read on.

7. ☐ Take your petri dish over to the ultraviolet light box. Place the dish on the box and close the lid. Turn on the light. What do you see?

The chlorophyll will glow (fluoresce) with a reddish color. Intact leaves *don't* fluoresce. The fluorescence that you see is energy being given off by chlorophyll in the form of light. This energy would normally be passed on to other compounds in the leaf to be used in the fixation of carbon dioxide.

Questions to Consider

Which colors (wavelengths) of light are mainly used in photosynthesis?

Why are leaves green?

Imagine you are standing on the floor of a dense tropical forest. High above you is a dense canopy of leaves. What color is the light around you? Why is it that color? (Bear in mind that your brain will "correct" for the appearance of the light – it'll look "normal" to you.)

Believe it or not, forest floors typically have relatively few plants. In terms of light, why is this so?

Activity 3: Storage Products of Photosynthesis

Within chloroplasts, the light energy harvested by chlorophyll is channeled into the formation of high-energy bonds in the carbon compound glucose. As more and more glucose is produced, it is stored as a polymer (a *polymer* is a chain of smaller units), called starch. Starch molecules are mostly insoluble in water and resist degradation except by specific enzymes, making starch a reliable way to store glucose. Plants can later break down their stores of starch into glucose for respiration. You could predict that the most starch will be stored in the parts of a leaf that receive the greatest exposure to light.

Procedure: Where on Leaves are Starches Stored?

Your instructor *may* have started some of the procedures for you and will let you know which ones are left for you to do.

1. ☐ Work in pairs or small groups.

2. ☐ Obtain a leaf from the cooler.

3. ☐ Place the leaf in the large beaker of 80% ethanol on the hotplate. More than one group of students can use the same beaker. Be sure to place the leaf in the ethanol BEFORE turning on the hot plate.

4. ☐ Boil the leaf in the alcohol until ***all*** the green color is gone from the leaf, including from the veins. The leaf should look "whitish."

5. ☐ Use tongs to remove the leaf from the hot alcohol. Place the leaf in a petri dish and completely cover/soak it with iodine solution from the dropper bottle at your lab table. Don't be stingy with the iodine and ask for more if your bottle gets low.

   When starch is exposed to iodine, a dark blue-black color develops.

6. ☐ Allow time for the color to develop.

7. ☐ Draw a picture of your leaf, including the pattern of light and dark areas, in the space below. Indicate with a label and arrow which parts of the leaf were exposed to light and which were shaded.

8. ☐ CLEAN UP: Wash and dry the beaker and plates and return to the box at your lab table.

Questions to Consider

How did a picture end up on your leaf?

What happens to the starch content of a plant's leaves after several cloudy days?

Which would be a better time to harvest and eat raspberries: after a rainy spell or after several sunny, dry days? Explain.

Activity 4 (Activity 1 continued): Measure Dissolved Gasses: Completing the Photosynthesis and Respiration Experiment

Once the sixty minutes are up from Activity 1, you will test the water in the four beakers for its dissolved carbon dioxide (CO_2) and oxygen (O_2) content. You will not be directly measuring CO_2 and O_2, however. Rather, you'll be measuring compounds directly related to the amounts present. This is a common practice in science.

Procedure 1: Testing for Dissolved Carbon Dioxide

The following procedure is based on the fact that when CO_2 dissolves in water it forms carbonic acid ($H_2O + CO_2 \rightarrow H_2CO_3$), so the solution becomes more acidic. (The bubbles in soap are CO_2 and, therefore, soda is slightly acidic.) If you add NaOH (sodium hydroxide, a base) to the water, it begins to neutralize the carbonic acid. By counting the number of drops of a NaOH solution it takes to neutralize the acid, you can tell you how much acid (and therefore CO_2) was in the water. *More NaOH means more CO_2!*

1. ☐ Label four 50 ml flasks with the four treatments: Control, Light, Dark, and Fish.

2. ☐ Arrange the flasks on a sheet of white paper (paper toweling works fine).

3. ☐ From **each** of the larger (100 ml) beakers, slowly and carefully pour **exactly** 20 ml of water into the *correspondingly labeled* 50 ml flask. Always keep the original "Dark" treatment beaker (the one with the elodea still in it) out of the light. When pouring water from the beaker with the fish in it, just pull back a bit of the plastic wrap from the spout area of the beaker -- enough so that you can pour out some of the water without the fish being able to slip out through the opening.

4. ☐ Test for dissolved carbon dioxide:

    (a) ☐ Add four drops of phenolphthalein to each flask. Swirl **gently** to mix.

    Phenolphthalein is a pH indicator that is colorless when a solution is acidic and pink when a solution is basic. Thus, a solution with lots of dissolved CO_2 is colorless, and a solution with little CO_2 is light pink.

    (b) ☐ If some of your solutions have already turned pink, determine which one is the deepest pink. **DO NOT** add any NaOH to this solution. Record it as having zero drops of NaOH added. Go to step "c" and skip step "d".

    ☐ If none of your solutions tuned pink, go to step "d" and skip step "c".

(c) ☐ Use an eye dropper to add one drop of 5 mM NaOH (sodium hydroxide) to the water in each of the other flasks. Swirl gently and slowly until the solution is mixed. Add more NaOH, **one drop at a time, counting the drops**, until the color matches that of the flask determined in step "b" above. Be sure to gently swirl the solution after each drop. Go to step "e".

(d) ☐ Use an eye dropper to add one drop of 5 mM NaOH (sodium hydroxide) to the water in each of the flasks. Swirl gently and slowly until the solution is mixed. Add more NaOH, **one drop at a time, counting the drops**, until the water just turns pink. Go to step "e".

(e) ☐ Record the number of drops of NaOH it took to neutralize the carbonic acid (dissolved CO_2) in Table 1 on the previous page.

Table 1. Dissolved oxygen and carbon dioxide concentrations in water from four treatments. The more drops added, the higher the concentration of O_2 or CO_2.

Treatment	**Which occurred?** **Photosynthesis? Respiration? Both?**	**Carbon Dioxide** **(Number of drops of NaOH)**	**Dissolved Oxygen** **(ppm)**
Control			
Plant in Light			
Plant in Dark			
Fish			

Procedure 2: Testing for Dissolved Oxygen

Your instructor will inform you which method of oxygen testing (A or B) you will be using today. Follow the directions for that oxygen method only.

Oxygen Testing Method A: CHEMets® Kit

1. ☐ Using tape, label four 25ml or 50 ml beakers with the four treatments: Control, Light, Dark, and Fish.

2. ☐ Slowly and carefully pour water from each treatment down the side of the appropriate beaker up to the 25mL mark. You do not want to introduce any air into the solution by pouring too fast or by making bubbles.

3. ☐ Open up the CHEMets® Kit at your table and take out 4 ampules (small glass tubes filled with liquid) from the box. **Be sure to close the kit** to limit the contents exposure to light.

4. ☐ With a permanent marker, gently label each ampule with one of the four treatment conditions.

5. ☐ Obtain protective eyewear and latex gloves for the person who will be breaking the ampule.

6. ☐ Place each ampule, tip side down, into the beaker that matches its treatment label.

7. ☐ While the ampule is immersed in the sample, gently but firmly snap the tip of the ampule (near the "ring" on the tip) against the side of the beaker and leave the ampule in the beaker. The ampule will fill on its own and will leave a small bubble to allow mixing.

8. ☐ Mix the contents of the ampule by inverting it several times, allowing the bubble to go from one end of the tube to the other. Dry the ampule off and set aside (out of the beaker) for 1-2 minutes so that the color can form.

9. ☐ Repeat steps 7-8 for the rest of the treatment beakers and ampules.

10. ☐ To determine the dissolved oxygen content of each treatment, open the CHEMets® Kit so that you can see the different colored tubes (these are known as color standards). Place an ampule between the color standards moving it from left to right until the best color match is found. If the color of the ampule is between two standards, you may estimate the concentration as being halfway between the two standards. For example, if the ampule matches between 2 ppm and 3 pmm one can estimate that the ampule is 2.5 ppm. Record this number in Table 1.

11. ☐ Repeat Step 10 with all other treatment ampules. **In general, the higher the number the more oxygen that was in the sample.** *Close the kit box when done using the color standards.*

12. ☐ Using the "Dissolved Oxygen Tolerance in Fish" diagram determine how your fish is feeling today, after being through the experiment.

My fish is feeling ____.

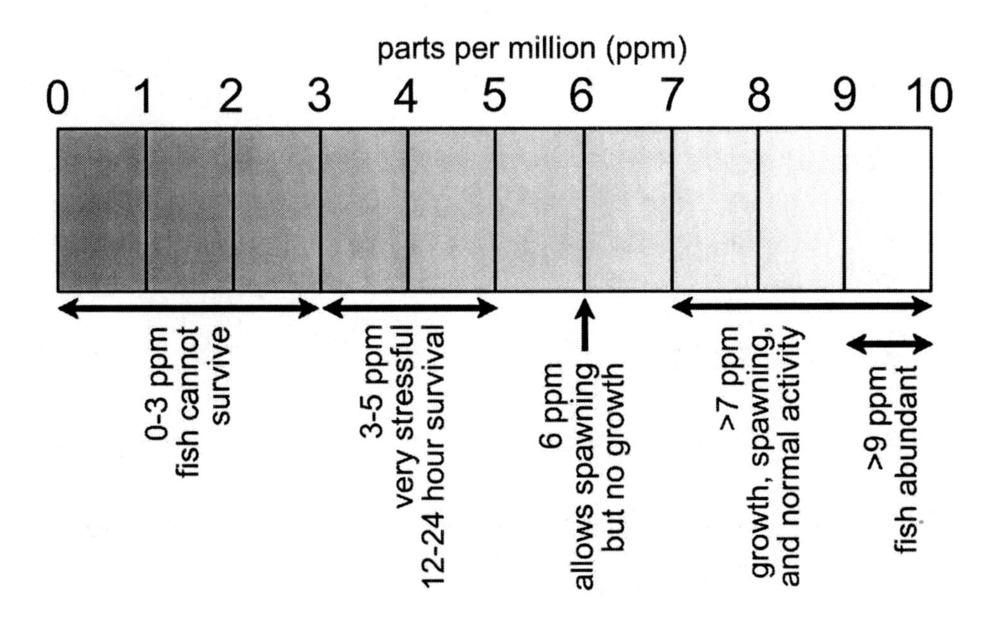

13. ☐ Pour the remaining liquid from the beaker the ampule was broken in into the sink – being careful not to let the broken tip go into the sink. Dump the glass tip into the spot designated by your instructor.

14. ☐ When done with the ampule, your instructor will show you where to properly dispose of them.

15. ☐ Once everyone in your group is done using the "fish water" (that is both the O_2 and CO_2 beakers have been filled), release the fish back into the tank.

16. ☐ CLEAN UP: The solutions in all the beakers can go down the drain. Rinse all glassware thoroughly, drain, then and place the stoppers back in the correct flasks. Return glassware to their correct places on your lab bench.

Oxygen Testing Method B

If you completed Oxygen Testing Method A, skip this section for Oxygen Testing Method B

1. ☐ Label four 25-ml flasks with the four treatments: Control, Light, Dark, Fish.

2. ☐ Slowly and carefully pour water from each treatment down the side of the appropriate 25-ml flask up to the 25 ml mark. You do not want to introduce any air into the solution by pouring too fast or by making bubbles. Always keep the "Dark" treatment out of the light.

3. ☐ Add 8 drops of Reagent #1 to each flask.

4. ☐ Add 5 drops of Reagent #2 to each flask. Close the flasks with a stopper.

5. ☐ ***Gently*** invert the flasks to mix the reagents. You will notice a fluffy precipitate form.

   The precipitate is formed when manganous sulfate reacts with oxygen to form manganese oxide (MnO_2), which is insoluble in water.

6. ☐ Let the precipitate settle until at least the top half of the solution is clear.

   This ensures that all of the oxygen dissolved in the water is *fixed* as MnO_2.

7. ☐ Add a small, level scoop of sulfamic acid to each flask. Close the flasks and gently invert until all of the sulfamic acid is dissolved.

   The solution will turn yellow or brownish. The darker the color, the more oxygen that was dissolved in the original water. Sulfamic acid causes MnO_2 and potassium iodide (KI, from Reagent #2) to break down in such a way that water and iodine (I_2) are formed. One I_2 is formed for every O_2 that was originally in the water. The change in color is due to the presence of I_2. The entire reaction is: $MnO_2 + 4H^+ + 2I^- \rightarrow Mn^{2+} + I_2 + 2H_2O$. Don't worry, you are not expected to remember this equation!

8. ☐ Add a drop of sodium thiosulfate to each flask. Close and mix gently but well.

9. ☐ Repeat the last step for each flask until the solution just turns <u>completely</u> clear. Keep track of the number of drops that were needed to turn each solution clear. Record this number in Table 1.

More drops of sodium thiosulfate = more oxygen in the original sample. Sodium thiosulfate causes iodine (I_2) to break into two iodide ($2I^-$) ions. Iodide is colorless. So, by knowing how many drops of sodium thiosulfate it takes to change all of the iodine into iodide, you also have a measure of the amount of oxygen that was in the original sample. Remember, each oxygen molecule was responsible for the formation of a single iodine molecule.

10. ☐ **Wash your hands!** Some of the reagents are slightly toxic and irritating at the concentrations we are using.

11. ☐ CLEAN UP.

Questions to Consider

Refer to your predictions on page 3. Are your predictions supported by the data in Table 1? Explain.

Which treatment had the highest dissolved oxygen content? Why?

Was the dissolved oxygen content of the beaker containing a plant in the dark less than that of the control? Explain.

Which treatment had the highest dissolved CO_2 content? What process resulted in the release of CO_2 into the water?

Did the "plant in the dark" or the "plant in the light" treatment result in a lower CO_2 content? Why would you expect lower CO_2 content with a plant in the light?

Explain the relationship between the number of drops of NaOH you added and the amount of CO_2 in the water.

What would happen to the dissolved oxygen and CO_2 content of the water if you put both a plant and a fish in a beaker in the light?

Consider a piece of wood. Where did its mass come from, specifically **carbon** (C), its main component? Air? Water? Soil?

Name: ________________________________ Instructor's Name: ______________

Normal Lab Day and Time: ________________

Natural World Lab Response Sheet

Lab 3: Atmosphere and Energy

1. What was the main idea behind today's lab?

2. Explain how you measured the amount of CO_2 in your water samples. Be sure to include both chemicals used.

3. Which plant would have <u>more</u> photosynthesis occurring...one exposed to a light bulb that gives off (emits) only blue wavelengths or a light bulb that emits only yellow wavelengths? Why?

4. Rank today's lab from 0 (poor) to 10 (excellent). *Why?*

LAB 4: ESTIMATING BIODIVERSITY: THE SPECIES-AREA RELATIONSHIP

A *biological community* is the sum of all of the organisms that inhabit an area – the plants, animals, fungi, and microorganisms. Different communities, of course, each have a different number of species. For example, the community called "your backyard lawn" obviously has fewer species than does an equal-area community of tropical rain forest. But just how many species does an area have? You could painstakingly count all the species in the entire area. But in nature this is nearly impossible to do for a large area. Ecologists typically must settle for studying a *sample*, a smaller representative of a larger area, in order to estimate the number of species in a larger area.

Ecologists have known for a long time that increasingly larger areas have increasingly more species. In other words, if a square mile has 200 species in it, two square miles has *more* than this number, but not twice as many. This is known as the *species-area relationship*. This relationship, however, is not a linear one. Rather, it forms a curve of a well-known shape. (If you want to know, it is called a Power Curve of the form $S = cA^z$, where S is the number of species, A is area, and c and z are constants used to fit the curve to the data).

Why would larger areas have more species than smaller areas? Come up with some reasons and write them below.

The total number of species found in an area is referred to as *richness*, although it is commonly called *diversity* or *biodiversity* in newspapers and magazines. For ecologists, however, diversity (biodiversity) actually includes two components, richness and *evenness*. Evenness is a measure of how equivalent in total number of individuals each species is. For example, a community of three species each with 100 individuals is more *diverse* than a community of three species in which one species has 298 individuals and the others have only one individual.

In this lab, you will be a member of a team of ecologists charged with estimating the diversity of a community called Baker's Forest. Species in this hypothetical community are represented by different objects. Notice that some species are very abundant while others are rare. In nature, a few species are usually very abundant, most have intermediate numbers, and some are very rare. The proportions of different "species" in your hypothetical community have been constructed to mimic those of real communities – your community, however, has far fewer species than a typical natural community.

You will be using a sample, or *plot*, to estimate the richness of the entire Baker's Forest. The figure below illustrates the nested nature of the plots you will be using. Notice that small subplots are nested into larger and larger plots. This enables the ecologist to estimate the richness of different sized areas. The largest plot is 16 meters on a side, so it actually examines the species in a 256 m^2 area. A plot divided this way and of this size is called a *relevé plot*. Your plot is scaled to be about 2/5 of the total area of Baker's Forest. (It's a very small community.)

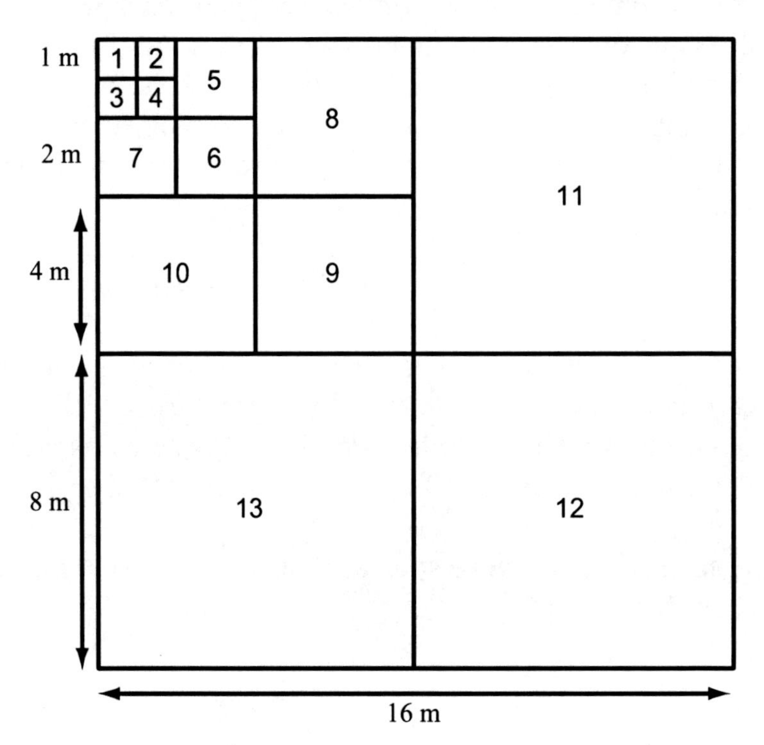

Diagram of a Relevé Plot

Procedure:

1. ☐ Divide yourselves into two "field teams" per table. Each team has its own forest to examine.

2. ☐ Take the container of "species" and scatter them into the Baker's Forest. Mix them well until they are more or less evenly distributed. (*Please be careful*, as the number of individuals of each species has been carefully counted to mimic the relative numbers of individuals in real communities.)

3. ☐ You've been given a plastic overlay with the relevé plot printed on it. Place the overlay anywhere in the Forest. Try not to be biased in your placement! Ecologist try to do this randomly.

4. ☐ Starting with Sub-Plot 1 (see diagram on previous page), record all of the species in this plot. Objects of the same color, size, and shape are the same species. Use the Plot Data Form to record your result. Put an "X" in the appropriate box for Plot 1 and record a description of the species to help you keep track of species already encountered. Remember that each line on the table is for a different species. Record only one "species" per line.

5. ☐ Move on to Sub-Plot 2 but record only additional new species. Again, record your data on the Plot Data Form but only record new species with an "X". Notice that you now know the number of species in a 1 m^2 and in a 2 m^2 area. Continue until you have examined all 13 plots. Remember, you only want to know the number of new species found in each additional plot.

Plot Data Form													
Plot Number													
1	2	3	4	5	6	7	8	9	10	11	12	13	Description of Species

6. ☐ Use the data in the Plot Data Form to fill in the Species-Area Data Form below.

Species-Area Data Form				
Plot	Plot Area (m^2)	Cumulative Plot Area (m^2) (graph this)	# of New Species per Plot	Cumulative # Species (graph this)
	1	1		
	1	2		
	1	3		
	1	4		
	4	8		
	4	12		
	4	16		
	16	32		
	16	48		
	16	64		
	64	128		
	64	192		
	64	256		

7. ☐ Use the data in the Species-Area Data Form to plot (with a dot) the Cumulative Number of Species by Cumulative Plot Area on the graph on the next page.

8. ☐ Draw a smooth line through the dots on your graph. You do not need to "connect the dots." Just make the line as smooth as possible. You've now constructed a Species-Area Curve! Big deal. So what?

The "so what" part comes in when you want to know the *predicted* number of species in an area larger than your sample plot. Like . . . how many species are in Baker's Forest altogether? Another answer to the "so what" question is that these kinds of curves help ecologists predict the size of an area that must be preserved if most of the species are to be saved in the long run. (Note that we are not considering here the ability of organisms to migrate into or out of our community from neighboring communities.)

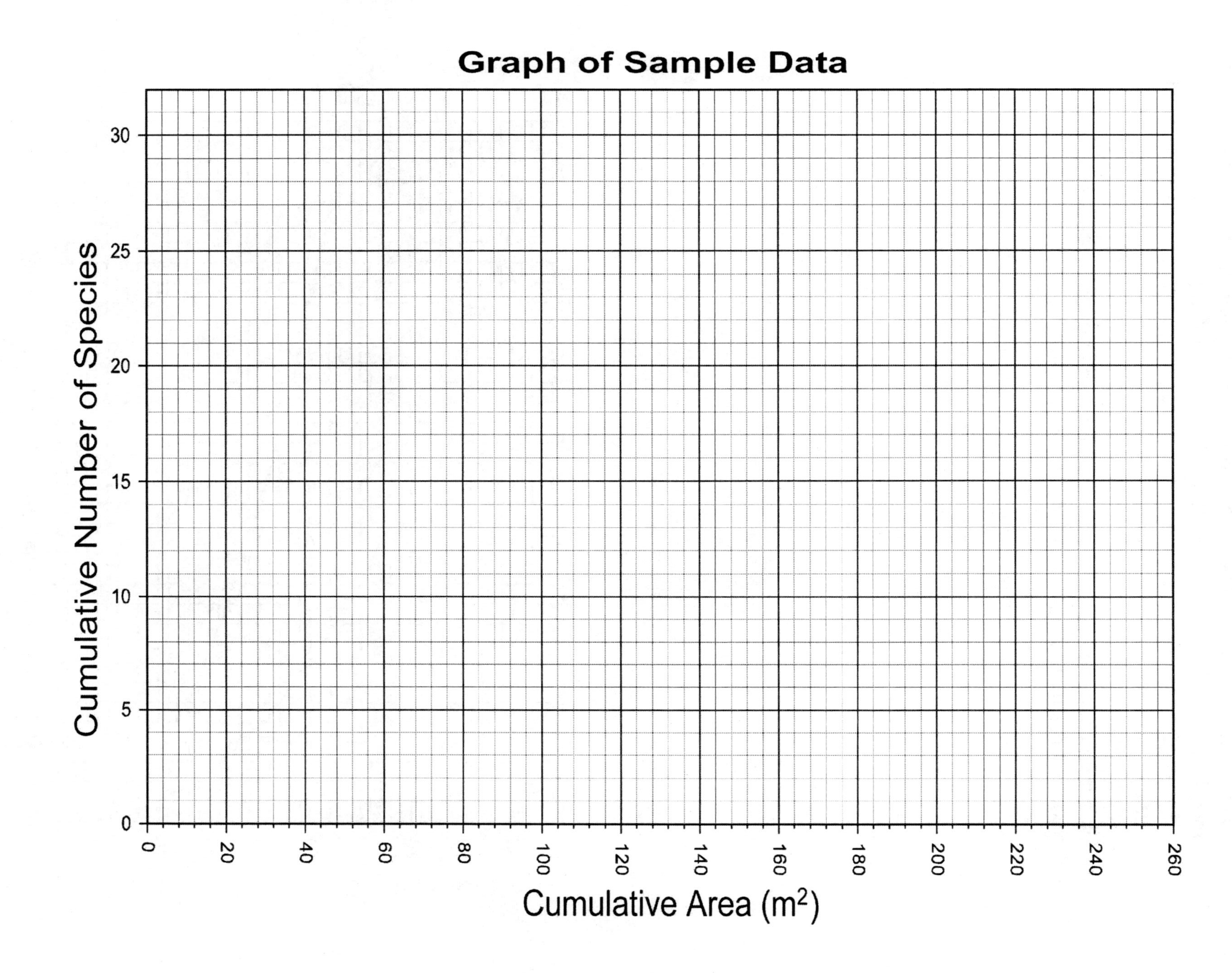
Graph of Sample Data
Cumulative Number of Species
30
25
20
15
10
5
0
0
20
40
60
80
100
120
140
160
180
200
220
240
260
Cumulative Area (m^2)

9. ☐ Refer to the pair of species-area curves in the graph below for two different communities. If each community is found to be 620 m^2 in area, which community, A or B, would need a larger area to be protected in order to save 90% of the species?

Which did you choose? Explain your choice giving the actual area of A and B you will need to preserve. Do the math and show your work.

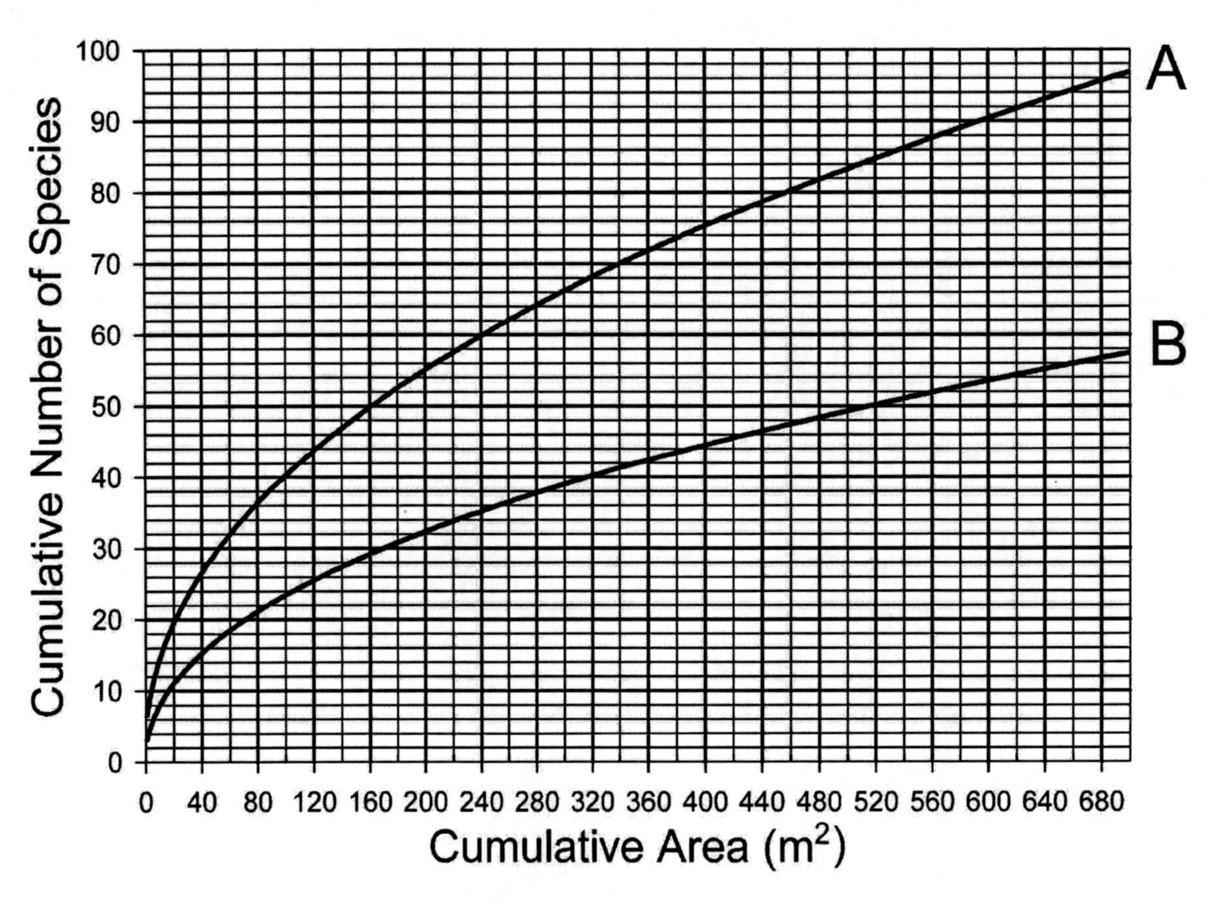

10. ☐ **Re-graph** your cumulative area and cumulative species data on the smaller scale graph on the next page. Look at your first graph, above, and observe the shape of the curve – especially towards the end of the graph. Use this version to *estimate* the slope (or curve) on your new population graph and extend the curve line to the end of the graph paper.

11. ☐ Use the population graph you just made to predict the total number of species in **Baker's Forest** (not the sample). You'll need to first measure the *inside* area of Baker's Forest (the baking tray). The scale of Baker's Forest is 1.25 cm = 1 m, so divide both length and width measured in cm by 1.25 to get meters. A ruler is provided. **Remember: Area equals length × width.**

    Area of Baker's Forest __________ m^2

    Total predicted number of species in Baker's forest (from your 2nd graph) __________

    How many more species are in the total predicted population estimate compared to the cummulative number of species found in the relevé sample?

    ____

12. ☐ Based on your curve, how much of Baker's Forest, in m^2, should be preserved if you want to save 85% of the total number of species? Give your reasoning and show your work directly on the graph.

13. ☐ Ask your instructor the actual number of species in Baker's Forest. ________

    How did your *estimate* of the total number of species in Baker's Forest compare with the *actual* number of species present? Provide at least one reason why your estimate of Baker's Forest's richness was or was not the same as its actual richness. (Think about sampling errors.)

Important!

Please put away the lab materials very carefully so as not to lose any parts. Thanks!

Graph of Population Estimate

Cumulative Number of Species: 0, 10, 20, 30, 40

Cumulative Area of the Whole Forest (m^2): 0, 50, 100, 150, 200, 250, 300, 350, 400, 450, 500, 550, 600, 650, 700

Question to Consider

You are a developer who would like to build a new office complex adjacent to and including a part of a 500 m^2 prairie remnant. Your development plan requires that 40% of the prairie remnant be destroyed to build the office complex. However, a county ordinance states that "...new developments can proceed only if the remaining area of any affected prairie remnants is sufficient to sustain 85% of the species that lived there prior to development...." After doing some field research, you create a Species-Area Curve of the prairie reserve. Based on the information from the graph below, will you be allowed to develop on the prairie reserve, or will you have to look for a new site? Be sure to show your work so that you can justify your answer.

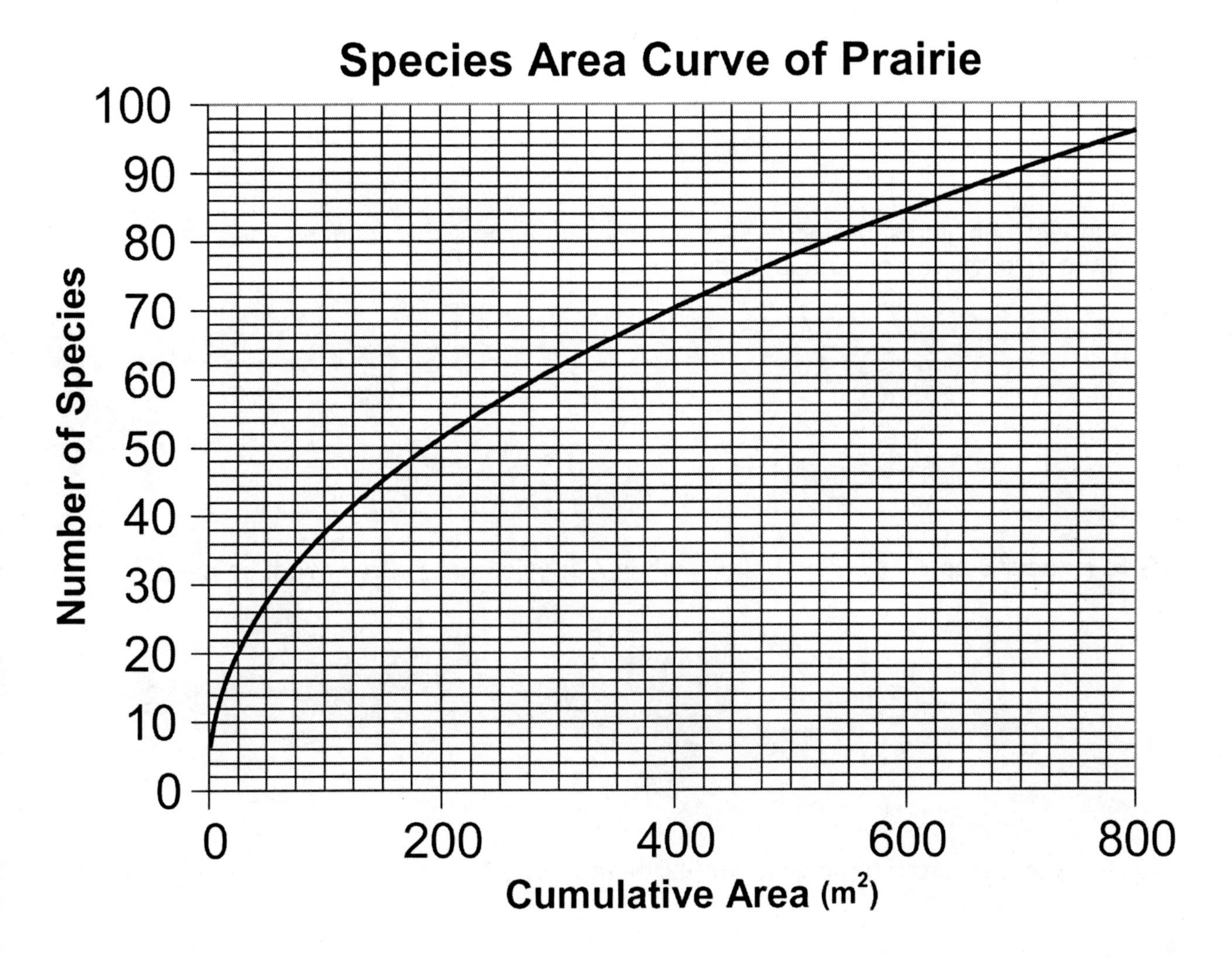

Name: ______________________________ Instructor's Name: _____________

Normal Lab Day and Time: _______________

Natural World Lab Response Sheet

Lab 4: Species Area

1. What was the main idea behind today's lab?

2. Describe why biologist prefer integrating two measurements (richness and eveness) to understand diversity rather than just using one.

3. Give a specific application example of when a species-area curve might be used.

4. Rank today's lab from 0 (poor) to 10 (excellent). *Why?*

LAB 5: DNA DIVERSITY AND ITS MEASUREMENT

In this lab you will explore some of the techniques used to study the *genetic diversity* present in humans. Upon completion, you will have used some of the DNA techniques commonly used by biologists to examine genetic variation within populations of organisms and among species. These are, incidentally, the same techniques forensic scientists working for police departments and the FBI use to identify or eliminate suspects in crimes.

Each human cell contains 6.4 billion base pairs of DNA — about 6 feet of DNA if the molecules are unwound and stretched end to end! Portions of this DNA code for about 24,000 protein-coding genes while some of the remainder codes for various kinds of RNA that don't lead to protein formation. (RNA is a molecule made from and similar to DNA which mainly directs the construction of proteins. Some RNAs have other functions that don't involve protein formation, like determining when to make a particular protein). Mostly, however, our DNA is non-coding. Human DNA is more than 99.9% identical among all people. In protein-coding genes, the DNA of humans is 98.5% the same as that in chimpanzees! Obviously that 1.5% difference means a lot. Likewise the <0.1% of DNA that differs between individual people matters quite a bit too. Look around the room – we're all different.

Normally, humans have 23 pairs of chromosomes (structures in the cell nucleus made up of DNA and associated proteins). Therefore, we have a total of 46 chromosomes. We have two copies of chromosome 1, two copies of chromosome 2, and so on. You get one copy of each chromosome from your mother and the other from your father – in other words, half of your DNA comes from your mother's egg and the other half comes from one of your father's sperm cells. Chromosomes carry the DNA instructions to make the proteins and regulatory molecules used to build all of our parts – fingers, heart, blood cells, eye color, hormones, etc. The instructions for making a particular trait or characteristic are found along a length of DNA known as a *gene*. Within a gene, the "instructions" actually consist of the particular order of the four nucleotide bases (A adenine, C cytosine, G guanine, and T thymine). The genes found on Chromosome 1 from your mom are the same, and in the same location (*locus*, plural *loci*), as the genes on Chromosome 1 from your dad. So, you have two copies of every gene (except some that are found only on the X and Y sex chromosomes). Genes themselves usually come in different versions; these different versions are called *alleles*. If you have two copies of the same allele of a gene you are said to be *homozygous*. If your two alleles are different, you are *heterozygous* for that gene. For instance, we all have two copies of the gene for the protein that determines blood type. If the gene from mom has the Type A allele, and the gene from dad has the Type B allele, then the person who has these two alleles will have the AB Blood Type (and has a heterozygous genetic combination for the trait). Each of our different chromosomes contains different genes (with their own versions, or alleles). For example, the Blood Type gene is located on Chromosome 9, but the gene involved with some types of freckles, the MC1R gene, is found on Chromosome 16.

DNA profile analysis focuses on extremely variable regions of DNA that are known to contain differences among people. This highly variable DNA is almost entirely found in our non-coding DNA. Because non-coding DNA doesn't code for anything useful, it is free to accumulate mutations with no harm to the individual. One such form of variable DNA is called a VNTR locus – short for Variable Number of Tandem Repeats. It seems that most, maybe all, eukaryotic organisms possess VNTR loci.

In this lab you will examine the D1S80 locus, one of the VNTR loci. D1S80 is located on human Chromosome 1 and is composed of small repeating units of DNA. Variability in the number of repeated units is the basis for DNA fingerprint analysis. Twenty-nine different *alleles* (versions of a "gene") have been identified and about 86% of people are heterozygous at the D1S80 locus. To begin, the Figure below illustrates the structure and makeup of a chromosome. Your lab instructor will explain it.

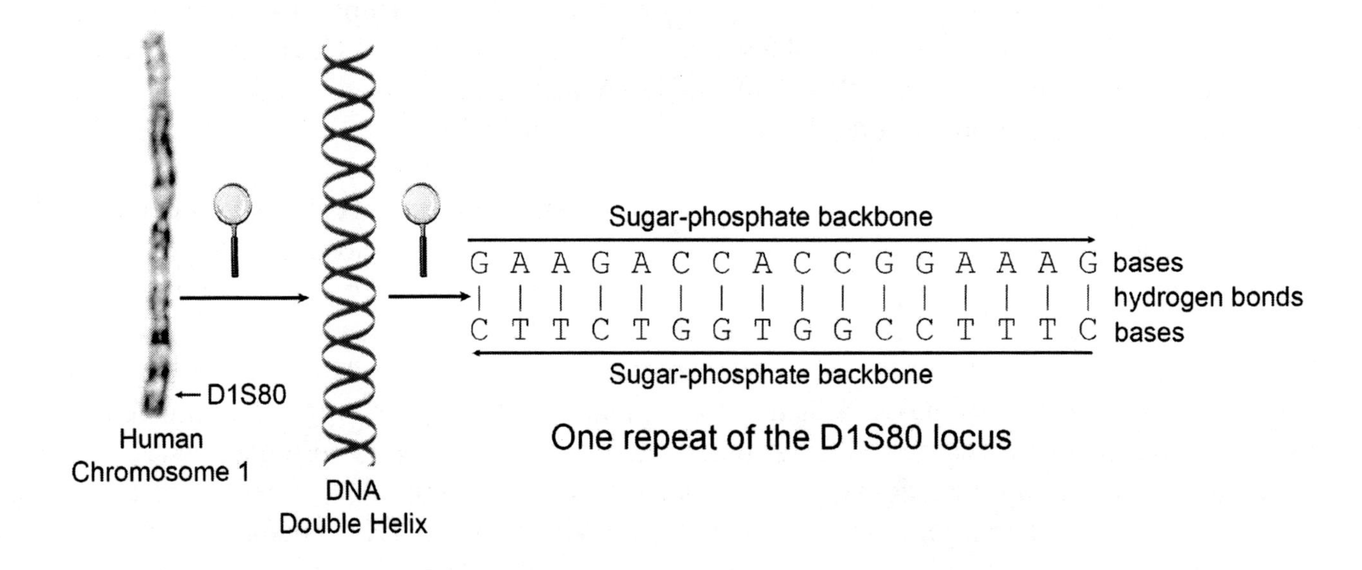

The following Figure illustrates the structure of the D1S80 genetic locus from a hypothetical person. You instructor will go over the details.

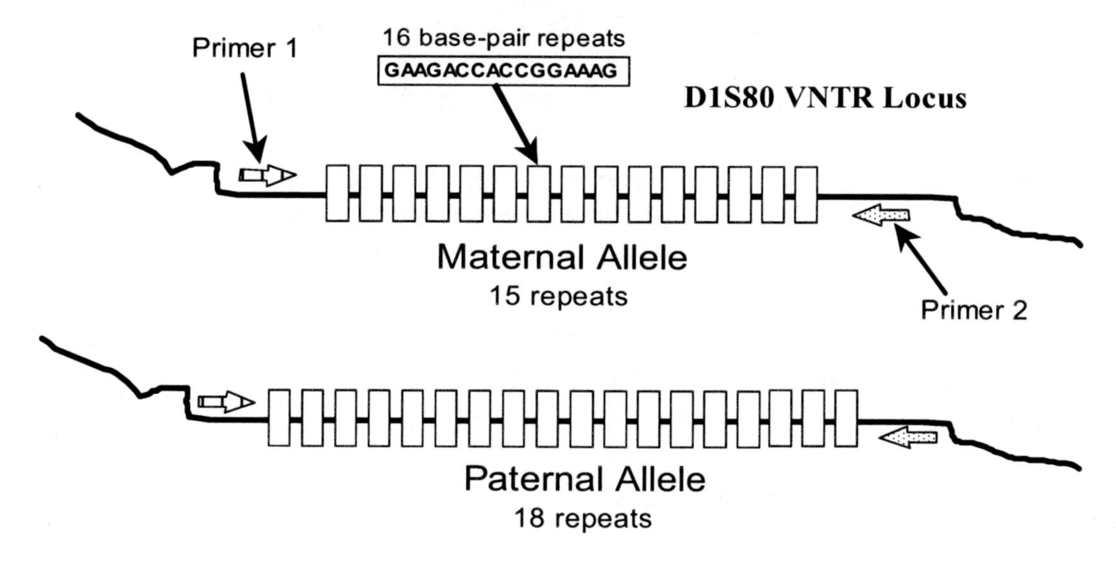

Lab 1 – Week One

Activity 1: DNA Isolation from Cheek Cells

DNA is locked up inside an organism's cells. Most of this DNA is contained in the linear chromosomes of the cell nucleus. A much smaller amount is also present as circular molecules inside the organelles called mitochondria. In addition to mitochondrial DNA, plants also have circular DNA molecules in their chloroplasts. In order to isolate DNA, we will need to rupture the cells and the cellular organelles. We'll simply place the cells in boiling water until they rupture. Once ruptured, cells will not only release their DNA, they'll also release compounds that can damage the DNA and that can interfere with further experiments using the DNA. A compound called Chelex (pronounced kee-lex) will be used to bind up some of the undesirable compounds. Turns out that cheek cells are a ready source of DNA – and a lot more user-friendly than having to draw some of your blood. Also, in mammals red blood cells have no DNA because they lack mitochondria and a nucleus. However, DNA can be extracted from white blood cells which do have a nucleus and mitochondria. Your instructor will go over how to use a pipettor, which is used to measure accurately small amounts of liquids.

Procedure 1: Extract your DNA

1. ☐ Label the side of a 1.5 ml flip-cap tube with your initials. Use the large pipettor to add 1000 µl (1 ml) of salt solution to the tube. (You can share pipette tips ONLY at this step. From now on, DO NOT share pipette tips.). Note that µl stands for *microliter*, one millionth of a liter.

   The salt solution is used to wash the cells. It has the same concentration of salt that your body does so that the cells do not rupture while being washed.

2. ☐ Vigorously and firmly scrape the inside of your cheek with a sterile toothpick.

3. ☐ Spin the toothpick between your fingers into the tube containing salt solution.

4. ☐ Vigorously mix the sample for 15 seconds using the vortex machine.

5. ☐ Place your tube in the centrifuge and spin at 8,000 rpm (revolutions per minute) for 1½ minutes. Carefully and gently remove your tube from the centrifuge and keep it in an upright position by placing it in a tube rack.

6. ☐ Carefully pour out the liquid into the sink or into a waste container being careful not to lose the pellet of cheek cells in the bottom of the tube. Gently shake to remove all liquid, again, being careful not to lose the cell pellet.

   The pellet consists of washed cheek cells.

7. ☐ **Important**: Thoroughly mix the Chelex solution just before you use it. Use the large pipettor to add 300 µl of Chelex solution to the sample in the tube.

   Chelex will bind up and remove undesirable compounds once the cheek cells are ruptured.

8. ☐ Vigorously mix your sample for 10 seconds.

9. ☐ Poke a small hole in the cap to keep the lid from popping open when the tube is placed in boiling water.

10. ☐ Boil in a water bath for 8 minutes.

    Boiling the cheek cells causes them to rupture and release their DNA into solution. It also deactivates most of the enzymes released from the cells.

11. ☐ Let cool for a minute, then vigorously mix the sample for 10 seconds. Spin in the centrifuge at maximum speed for 3 minutes. Carefully remove the tube from the centrifuge and set aside in an upright position in a tube rack.

    The pelet at bottom of the tube contains the Chelex and extraneous cellular material. The liquid contains *your* DNA dissolved in water. DNA will not be visible in the liquid.

    Your DNA sample is now ready for Polymerase Chain Reaction (PCR) amplification.

Activity 2: PCR Amplification of the D1S80 VNTR Locus

You've now isolated some of your own DNA. But how can we *see* a particular gene like D1S80? There are so few copies of any one gene in your sample that it is impossible to see it. We need to make many additional copies of the gene – we need to *amplify* it. Amplification of genes, or parts of genes, is accomplished by using the Polymerase Chain Reaction (PCR), which constructs many millions of copies of a gene. PCR uses an enzyme (DNA polymerase) to make copies of a particular section of DNA. New DNA molecules are built from the nucleotide building blocks A, C, G, and T. *Primers* are used that "recognize" a particular region of DNA. You instructor will go over the principle of the PCR. Refer to the Figure below.

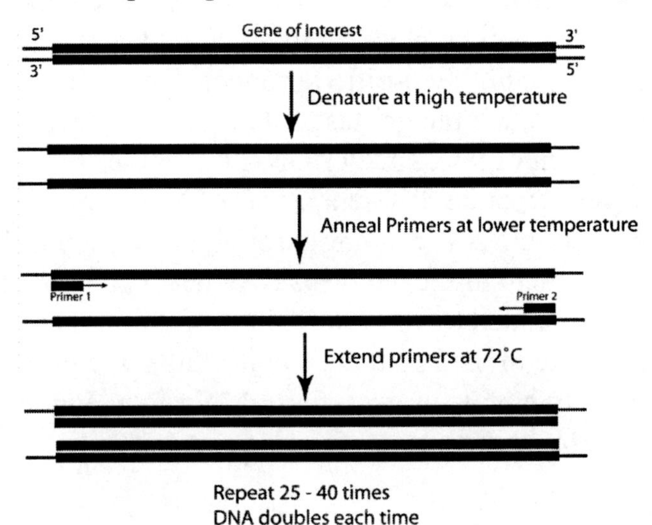

Procedure 2: Set Up the PCR

1. ☐ Without jiggling your tube, take **20 ml** of YOUR DNA solution from the TOP of the liquid from your tube and place it in the correct spot on PCR tray at the front of the classroom (your instructor will help you find the location on the tray).

2. ☐ Place your tube in the big blue box and look at the number underneath.

3. ☐ On the sheet provided by the instructor write down you name and PCR Tray spot so you can find your sample next week. Write this information below too.

   PCR Tray Location _______________

   Later on Reaction Mix will be added to your sample and it will then be processed through the PCR Reaction.

The Reaction Mix contains the enzyme DNA polymerase (that builds DNA molecules), the nucleotides A, C, G, and T (which are the building blocks of DNA), some salts needed by the polymerase, and a pH buffer to maintain a constant pH in the solution.

Your instructor will put your sample into a *thermocycler* machine and will run the machine for you. The next time lab meets we will "run" the samples in a gel so that the DNA pattern can be observed.

Lab 2 – Week Two

Activity 3: Gel Electrophoresis

PCR produces so many copies of a particular region of DNA that variants of the region can be seen if separated in an electric current and stained. DNA is a negatively charged molecule and will, therefore, move in an electric field from the negative end to the positive end. Your PCR amplified DNA will be placed into a *gel*, which is much like dense gelatin, only ours are made of *agarose*, a highly purified product of seaweed. In an electric field, small DNA molecules move through the gel faster than do large fragments. That's the same as saying that in a given amount of time smaller DNA molecules move farther than do larger molecules. We can, therefore, separate different sizes of DNA molecules in a processes called *electrophoresis*. As your DNA is running through the gel it will be stained by the dye ethidium bromide which is present in the gel. This dye, when combined with DNA, fluoresces brightly when illuminated by ultraviolet light. Ethidium bromides is a very effective stain. You will be able to see as little as 4 billionths of a gram of DNA! The Figure below shows diagrammatically how your gel will look. Known lengths of PCR products for D1S80 range from 238 to 1262 base pairs.

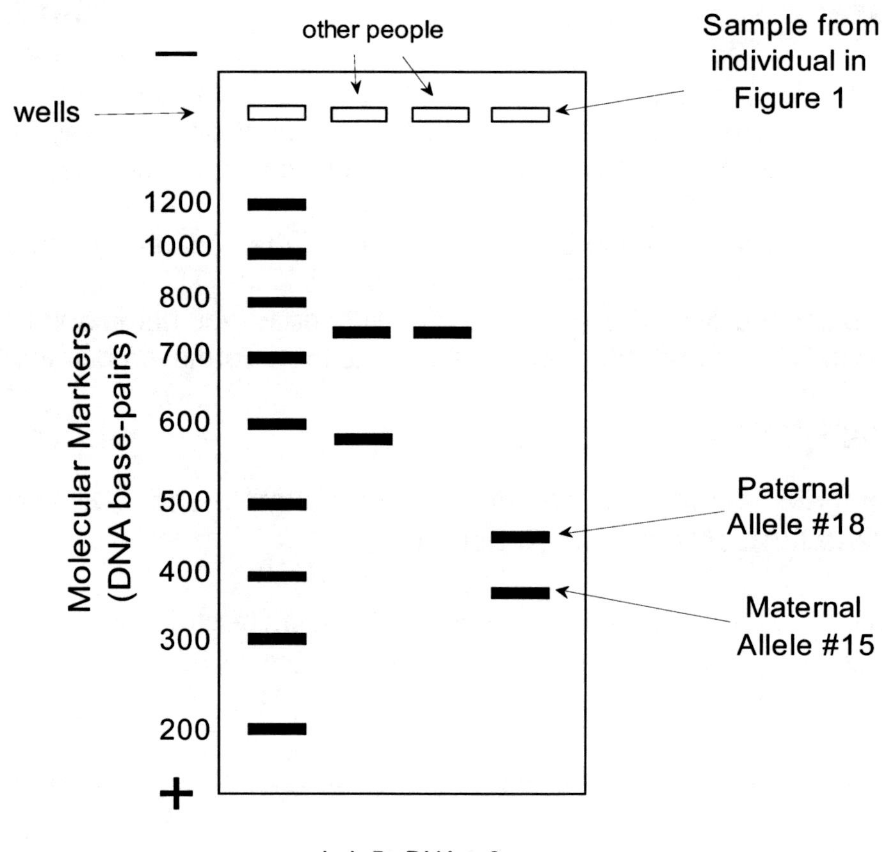

Recall that DNA is a

double helix, so every A, C, G, or T (the bases) is paired with another. A piece of DNA 12 base pairs long looks like this, where the dotted lines are hydrogen bonds.

```
A-C-C-G-A-T-A-G-C-G-A-G
:  :  :  :  :  :  :  :  :  :  :  :
T-G-G-C-T-A-T-C-G-C-T-C
```

Procedure 3: Gel Electrophoresis

1. ☐ Your instructor has prepared agarose gel cartridges in which you will electrophorese your DNA samples.

   The gel is made from agarose, ethidium bromide, and Electrophoresis Buffer, which contains salts to allow the gel to conduct electricity and is at pH 8. The gel is also submerged in the Electrophoresis Buffer.

2. ☐ Have the instructor load 5 µl of molecular weight standard into a MIDDLE well of the gel. Record this on the gel sign-up sheet.

3. ☐ Using a new pipette tip, carefully load **5 µl** of your sample into the next available well of the gel. Insert your pipette tip all the way to the bottom of the tube before removing the sample. Be sure not to poke your pipette tip into the bottom or sides of the gel! Record where you placed your DNA on the gel sign up sheet.

4. ☐ Your instructor will *electrophorese* your sample.

If your DNA comes out, you may do Activity 4 on your own to determine your genotype. Everyone should answer the last *Question to Consider* for today's lab.

Activity 4: So How Rare *Is* My Genotype?

First, you'll need to determine your *genotype*. You got one D1S80 *allele* from your mother and one from your father. If these two alleles are different, you are *heterozygous* at the D1S80 locus. If both alleles are identical, you are *homozygous.* Most people are heterozygous for D1S80 alleles.

Your instructor will help you take a photograph of your gel so that you and your group can examine it closely.

Procedure 4: Interpreting the Gel Results

1. ☐ You will refer to the known DNA markers on the gel to determine the size of your two alleles. Measure the distance in millimeters (mm) each of the <u>DNA markers</u> moved in the gel by measuring from the front of the well to the center of the DNA band. Try to estimate to the nearest half millimeter. Record your results in the table below. Your instructor will let you know the sizes of the marker bands.

Size of Marker Band (# of base pairs)	Distance Marker Band Moved (nearest 0.5 mm)

2. Plot your results on the *semilog* graph on the next page. If you need help, be sure to ask your instructor. Your instructor will let you know the length (size) of each marker band in DNA base pairs (A, C, G, and T).

3. ☐ Connect the plotted points using a smooth line.

4. ☐ Now measure the distance each of <u>your alleles</u> moved in the gel and read off the size of your alleles from your graph based on the curve you drew. How far did each of your alleles move in the gel? Allele 1 ___mm Allele 2 ___mm

5. ☐ Consult the D1S80 VNTR Locus Allele Frequencies Table given two pages below to find which alleles you possess. Choose the band sizes that are closest to your estimated band sizes.

6. ☐ Indicate your genotype by writing each of your alleles separated by a slash. For example, if you have the 27 and 40 alleles, write this as 27/40. Write your genotype here: _____________ .

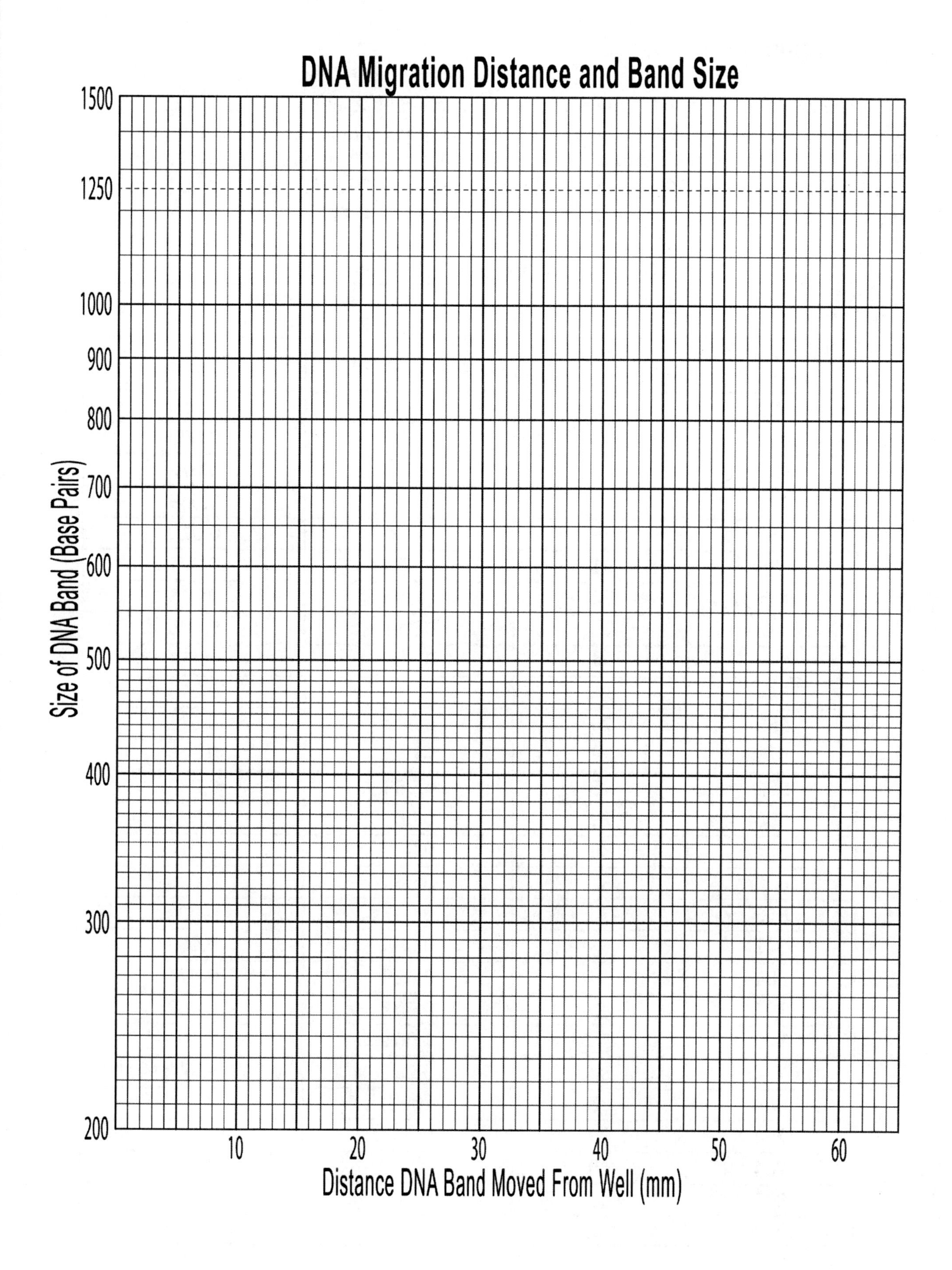
DNA Migration Distance and Band Size
1500
1250
1000
900
800
700
600
500
400
300
200
Size of DNA Band (Base Pairs)
10
20
30
40
50
60
Distance DNA Band Moved From Well (mm)

Your Genotype Probability

The probability of observing a particular genotype is found by using the table below. The table is based on data from people of all ethnic groups in the United States and in the World. Probability is just a way of indicating how common something is. For instance, if an allele has a frequency of 0.013, that allele is present in 1.3% of all people. Also, remember that you got one allele from your mother and one from your father.

D1S80 VNTR Locus Allele Frequencies

		Frequency (proportion of people with allele)	
Band Size (base pairs)	**Allele** (number of repeats)	**United States**	**World**
≤400	≤16	0.007	0.013
416	17	0.014	0.009
432	18	0.174	0.072
448	19	0.005	0.009
464	20	0.020	0.015
480	21	0.053	0.021
496	22	0.046	0.025
512	23	0.013	0.005
528	24	0.304	0.121
544	25	0.049	0.028
560	26	0.011	0.011
576	27	0.013	0.016
592	28	0.089	0.032
608	29	0.052	0.023
624	30	0.027	0.070
640	31	0.065	0.026
656	32	0.007	0.007
672	33	0.004	0.003
688	34	0.029	0.025
704	35	0.002	0.002
720	36	0.005	0.004
736	37	0.002	0.007 (>36 repeats)
752	38	0.0001	
768	39	0.003	
784	40	0.0001	
800	41	0.002	
>816	>41	0.004	

☐ To find the probability of *your* genotype following this example:

If you are heterozygous:

probability = 2 × (allele probability 1) × (allele probability 2)

If you are homozygous:

probability = (allele probability) × (allele probability)

For example, say you have a heterozygous genotype of 20/33. The probability of observing that genotype in the United States is (refer to the table above):

$2 \times 0.020 \times 0.004 = 0.00016$ In scientific notation that is 1.6×10^{-4}

If you have a homozygous genotype of 25/25 the probability of observing your genotype in the United States is:

$0.049 \times 0.049 = 0.00240$ In scientific notation that is 2.401×10^{-3}

To put this into context, there are about 311 million people in the United States. 311,000,000 × 0.00016 = 49,760 other people in the country are expected to have the 20/33 genotype and 311,000,000 × 0.002401 = 746,711 are expected to have the 25/25 genotype.

☐ Use the space below to calculate the U.S. and World probabilities of your genotype. Record your results here: U.S. ____________ World:____________

When detectives use DNA fingerprinting to identify or exclude a suspect they use several genetic markers. Because the probability of having a particular combination of several genetic markers is based on multiplying the probability of the pattern for each marker, the chances of convicting the wrong person are vanishingly small. Conversely, if the DNA says you are innocent, you are! For example, say four genetic markers are examined and the resulting genotypes have the following probabilities: 0.00242, 0.0253, 0.0057 and 0.0461. If we multiply those numbers together we get 0.000000016, or only 16 people out of a billion would be expected to have the same genotype. Given that there are about 6.8 billion people on earth, that translates to about 109 people on earth would have that genotype. The FBI regularly uses 16 makers, not four. Imagine those odds!

Questions to Consider:

What is *your* genotype (if your DNA worked out)?

What is the probability of finding this genotype in any given person in the World (if your DNA worked out)? Show your work.

Looking at the gel picture, did anyone else in the class have either of your alleles?

What was the most common allele in the class? Does this make sense in light of the table of probabilities you've been given? Explain.

Was anyone on your gel picture *homozygous* for the D1S80 locus? What was the frequency of *homozygosity* in the class? (Homozygotes divided by the total number of students.)

Suppose that you are suspected of a crime. DNA with your D1S80 genotype is found at the crime scene. Should a jury vote to convict you based on this evidence alone? Why or why not?

Name: ________________________________ Instructor's Name: ______________

Normal Lab Day and Time: ________________

Natural World Lab Response Sheet

Lab 5: DNA Part I – Week 1

1. What was the main goal for the DNA Part I Lab?

2. Would you expect to find more variation (diversity) among people in a part of our DNA that codes for a protein or in a part that is non-coding ("junk DNA"), such as a VNTR locus. Explain.

3. Describe an application of the study of genetic diversity. You may wish to give a specific example.

4. Which of the following is TRUE regarding pipetting technique:

   a) Push your thumb down to the first stop when the empty tip is in the liquid.

   b) Press the tip firmly against the bottom of a tube or bottle when attempting to draw up a measured quantity of liquid.

   c) Push your thumb down to the second "stop" only when expelling a measured quantity of liquid into a container.

   d) Once you have placed a sample into a new container, release your thumb, whether or not the tip is in the liquid.

Name: ________________________________ Instructor's Name: _____________

Normal Lab Day and Time: ________________

Natural World Lab Response Sheet

Lab 5: DNA Part II – Week 2

1. What was the main goal for the DNA **Part II** Lab?

2. Why do most people's D1S80 samples, a gene that has numerous alleles, show two DNA bands on the gel instead of just one?

3. We looked at the probability of anyone having your particular genotype at only the D1S80 locus. If we used several genetic markers (loci), what would happen to the probability of anyone having your specific genotype? Explain in general terms.

4. Rank the **entire DNA** lab from 0 (poor) to 10 (excellent). WHY did you choose this rating?

Lab 6: Reproducer – The Process of Natural Selection

As Charles Darwin surveyed the plants and animals along the coast of South America, one of the most significant places he visited was the Galápagos Islands. This cluster of volcanic islands lies on the equator 600 miles west of the coast of Ecuador in the Pacific ocean. Some islands contain populations of animals that are isolated from members of the same species on other islands. Darwin was struck by the variation in physical characteristics of animals living on different islands. Even more intriguing was the observation that different islands often had different, but closely related and similar, species. Some years later, while collecting specimens in the islands of Indonesia, Alfred Wallace came to the same conclusions when he observed variation in birds, butterflies, and other creatures among islands. Wallace and Darwin's island observations were fundamental to their independent development of the theory of natural selection – the main mechanism for the origin of new species – which was published concurrently by both of them in 1958.

The diversity of beak types among South American finches is a striking example of variation within and among species in the Galápagos Islands. Island finches share many characteristics with an unspecialized species found on mainland South America. Darwin wondered whether the island finches might have descended from a common ancestor from mainland finches that were accidentally blown off course while migrating. The mainland birds have large bills adapted to crushing hard seeds. The beaks of island finches are adapted to allow the birds to exploit a range of different food sources within and among different islands. Beaks now range from a massive seed-crushing bill to a very slender bill in an insect-eating species. There's even a vampire species that drinks the blood of other species!

Darwin's and Wallace's personal observations, coupled with concurrent developments in the science of geology – namely an earth much older than previously thought – led them to conclude that different organisms, like the thirteen or fourteen species of Galápagos finches, could have be derived over long periods of time from a common ancestor. The mechanism formulated for this process of speciation, called *natural selection*, can be summarized as follows:

- Individuals within a population vary in ways that can be inherited.

- Some individuals will have favorable traits that result in enhanced survival and reproductive success (i.e., enhanced *fitness*).

- More offspring are produced than can possibly survive and reproduce, which results in competition for resources.

- Over time, individuals having the favorable traits will make up an increasingly large proportion of the population.

Darwin also noticed that the males of some species had curious and seemingly deleterious ornamentations that were quite prominent. Why would something as big and costly as the colorful peacock tail exist – especially when it might make individuals more easily preyed upon? The answer lies with sexual attraction. Some ornamental characteristics are displays that are attractive to members of the opposite sex. The quality of the display may indicate health or "good genes" or they may help fight off other male competitors for access to fertile females (such as the enlarged claws of male fiddler crabs or the antlers of male elk). Even if there are survival costs to these traits, the costs of them are outweighed by increases in mating success (and thus more offspring). So, despite having a burdensome tail, peacocks reap the benefit of displaying it to peahens who can tell which males have good genes for survival and health from the number of eyespots on their tails. Gaudy ornamentation can be, for example, an indication of how many parasites a male carries. The selection of ornate or combative traits by one sex is known as *sexual selection*. Like natural selection, sexual selection also requires that individuals have heritable genetic differences in traits and that these differences influence the number of offspring that are produced and survive. However, the traits that sexual selection act on influence an organism's ability to attract and mate with members of the opposite sex or enable them to successfully compete for mates.

Activity 1: Modeling Selection

In this lab, your classmates will represent a population of a species of fish – Jemez Spring Guppies from New Mexico – that have slight variations of scale pigmentation. These variations in scale color are based on a carotenoid pigment (a compound similar to vitamin A). The goal of individuals will be to obtain enough food in order to survive and to produce as many offspring as possible (or in other words to have the highest *fitness* – a measure of how many offspring an individual can get into the future; said another way, a measure of the number of genes an individual can get into the future). Over the course of several seasons, you will keep track of the number of offspring that you produced each breeding season and the factors that influenced your ability to reproduce. You will also be keeping track of the number and variation of prey eaten at food sites and how much they reproduce.

Procedures:

1. ☐ You will each choose a ticket listing guppy parts from the "gene pool." You may use only the materials that are listed in your genetic makeup list to make the guppy. You have 10 minutes to use these materials to construct a "guppy" that you think will be able to pick up the available food source. When you are manufacturing your guppy take into consideration that you may use only one hand to operate it. (See "rules for using your fish" in Step 4.)

2. ☐ The guppies the class constructs represent the starting generation. Female fish lack, and male fish have carotenoid spots. In our model, the gene for carotenoid spots on the scales is carried on the Y, or male, chromosome – which is why

females fish do not have the spot. Record here the starting numbers of female guppies and the number of males in each carotenoid spot variation.

Females: Starting # of Female Guppies _______

Males: Starting # of Random Orange Spot Guppies _______

Starting # of Symmetrical Orange Spot Guppies _______

Starting # of Random Blue Spot Guppies _______

Starting # of Symmetrical Blue Spot Guppies _______

3. ☐ The food source, "aquatic worms," is in a plastic bag. Scatter the worms on your colored habitat. You will be foraging for this prey later on. Count the total number of individuals in each worm variation type for <u>all food sites on your lab bench</u> and record this in the **Start** column on the **Food Site Data** sheet at your bench.

Note: It will be important that you continually and <u>accurately</u> update this data sheet during the lab. Your instructor will be collecting this sheet at the end of class to help illustrate today's concepts.

4. ☐ Practice catching the worms with your guppy and place them in a paper or styrofoam cup.

Your goal is to eat as many worms as possible as this will impact the number of offspring you can have. Many factors can influence how many offspring an organism can produce, such as good health, lack of infections, and predator evasion. Having extra food energy leftover once energy requirements for survival are met allows for reproduction (and even extra offspring if you're a "super" forager). Guppies are live bearers – rather than laying eggs, they give birth to live, fully functioning, and independent fry (baby fish). Typically guppies have between 5 and 30 fry at a time, but this can range from 2–100.

Rules for Using Your Fish and Catching Prey:

(a) Do not use your hands to clean food from the mouth of the guppy.
(b) Only worms IN your cup count as food that you have eaten.
(c) You are only allowed to pick up one worm at a time.
(d) You may only use one hand to operate your guppy.
(e) You must pick up worms with the **tip** of your mouth.

5. ☐ When the timer says "Go," use your guppy to move as many worms as possible from the food sites into your cup. Stop when until the timer says "Stop." Remember that it is "legal" to stray to other food sites if your food source is depleted or not yielding you an *optimal* return (in other words too much work for the amount of food you get).

6. ☐ After the foraging time is over, record the results on the **Food Site Data** sheet at your bench. For **Generation 1** write down the number of surviving worms for each color variation (what worms are left on the food site). Keep the surviving worms on the food site.

7. ☐ Our worms reproduce asexually, and thus will produce offspring genetically identical to themselves. <u>Each surviving worm will have 2 "wormlets" every season</u>. Record the total number of new wormlets this season for each of the worm variations in the "# of New Offspring" column of Generation 1 (i.e., multiply the number of surviving worms of one variation by 2).

8. ☐ For each worm variation, add the # of surviving worms with the # of new offspring for Generation 1 and write this number in the "Total for Generation" row of the Generation 1 Column.

9. ☐ Now take out worms from the colored baggies that equal the number of **<u>new offspring</u>** you calculated for each of the worm variations. Randomly place these new wormlets on the food site with their parents. The next time you forage these new wormlets and their parents are fair game for food.

10. ☐ It is time to see how **YOU** did in the competition! Count up the prey in your cup. Write the number of total worms you caught in **Table 1** of your lab manual, under the "Season 1" column. Also, in the "Notes" section under the same column, write any comments about what helped or hindered your foraging yield. Place all eaten worms in the appropriate colored plastic baggies on your table for the worm color variation they are.

11. ☐ If you ate 15 worms in the allotted time, then you move on to the next level – mate attraction! If you caught less than 15 worms you have enough food energy to sustain yourself, but, sorry, no extra for reproduction. Think about what you can do next season to increase your food yield so you can reproduce. On Table 1 write a "zero" for "# of fry" in the "Season 1" column.

12. ☐ Now you must find a mate if you caught 15 or more worms! You are going to add your number of caught worms to that of your partners to get a total # of worms as a "couple." The higher the total number of worms the more fry (babies) your couple can have (see the table below). Remember the name of the game is to pass on your genes to the future population by having as many viable offspring as possible during your lifetime, so pick your mate carefully!

    Travel around the room advertising how much food you caught , or how fat you are, to attract an opposite sex mate. Remember, males have spots, females do not. In this model we can only mate with one partner each season, so if you cannot find a partner you are out of luck (even if you had enough food to reproduce)!

Male Jemez Spring guppies usually do a dynamic courtship "dance" to entice females to mate with them. This involves flashing opposite sides of their body back and fourth to the female. The more energetic the courtship behavior the "hotter" the female guppy thinks the male is. If you are having problems finding a mate you may want to try this out – at the very least it will be sure to get attention!

Total # of Worms of the "Couple"	Total # of Fry (Babies) Produced
30-39	2
40-49	6
50-59	10
60-69	16
70+	20

13. ☐ If you are lucky enough to a find and attract a mate, calculate the number of offspring you can have together using the table above. Write down the number of fry you had this season in Table 1 under the column Season 1. You instructor will poll the class to find out how many fry were born and what spot color the fathers had and will record the result in a table that you will copy down later.

14. ☐ Time to start again! However, this round should go faster since you know the general plan. We will be repeating steps 5–13 to do Generation 2 for the worms and breeding Season 2 for you. Be sure to record worm data and your own offspring data each season. We will continue to repeat these steps until your instructor announces that your life span is over. Listen carefully before each season, as you instructor may report changes in the environment or conditions along the way.

15. ☐ When your life span is over, please pass the **Food Site Data** sheets to your instructor. Your instructor will tally up the sheets and give the information to complete **Table 3**. Later your instructor will also give the data to fill in **Table 2**.

16. ☐ CLEAN UP: Once your life span is over, please return all colored worms to their appropriate bags. Gently dismantle your guppies and wrap a rubber band around your tongue depressors and return them to the correct box. Return other undamaged guppy parts to their containers too.

Table 1: Individual Food and Reproduction Tally

You will fill in this table while doing the activity.

	Season 1	Season 2	Season 3	Season 4	Season 5
# Worms You Ate					
Total # of Worms Eaten by You and Your "Mate"					
# of Fry					
Father's Coloration					
Notes					

Table 2: Offspring of All Guppies. Assume that half of each season's fry are male and half are female. Your Instructor will post these data for you to copy here.

# Offspring sired by...	Starting Population	Season 1	Season 2	Season 3	Season 4	Season 5
Males with random orange spots						
Males with symmetrical orange spots						
Males with random blue spots						
Males with symmetrical blue spots						

Questions to Consider

What was the total number of offspring for the highest reproducing male and female in the class?

Male: ______________ Female: ______________

What were some factors that influenced how much food you caught?

What were some factors that influenced how many offspring you or others had in each season?

Compare the starting numbers of male color phenotypes (Season 1) with the ending/last season variations. Which phenotypes increased and which decreased? What factors may have influenced any changes to number of phenotypes that occured?

If this color variation preference trend continued over even more time, what do you think would happen to the proportion of males with each color variation?

Give an example from our model of when color variation would be under the influence of sexual selection. Provide a different example of when it would be under the influence of natural selection.

Table 3: Number of Worms Present Across Generations

Your Instructor will post these data for you to copy here.

WORM DATA		Start	Generation 1	Generation 2	Generation 3	Generation 4
Black	Total for generation					
Speckled	Total for generation					
Gray	Total for generation					

Activity 2: Changes in Genotypes and Allele Frequencies

You will use your data to compare the proportions of different worm colors at the beginning and end of the activity and to calculate the frequency (percent) change in the alleles that determine color. A particular combination of alleles gives a *genotype*. In our example, an AA genotype gives a black worm, Aa gives a speckled worm, and aa gives a gray worm. Each genotype, then produces its own *phenotype* (physical attribute, like color). This is common. Perhaps in another class your were taught about dominant and recessive alleles. In those cases AA and Aa produce one phenotype and aa produces another – assuming that the A allele is dominant over the a allele.

Procedures:

1. ☐ Use your data in Table 3, to complete Table 4, below. First, fill in the number of each colored worm we had in our starting generation ("Start" column). Add up the black, speckled, and gray worm number to calculate the "Total Worms."

2. ☐ You need to determine the percent of each worm genotype so we can easily compare the groups. Use the following example to help you calculate the percentages of different worm color variations in the population:

$$\% \text{ Black} = \frac{\text{\# of Black Worms}}{\text{\# of Total Worms}} \times 100$$

where # Total Worms = # Black + # Speckled + # Gray

3. ☐ Now do the same for the "Ending" column, except use the values in "Generation 4" of Table 3.

Table 4: Quantity of Genotypes Across Generations

Worm Phenotype	Starting		Ending		% Change from Start to End
	#	%	#	%	
Black (AA)					
Speckled (Aa)					
Gray (aa)					
Total worms		100%		100%	

4. ☐ To calculate "Change from Start to End," subtract your Ending percent value from your Starting percent value for each worm color variation. Be sure to include the positive or negative sign of the calculated value. The percent change will tell you if that particular variation increased or decreased (and by how much) over the range of generations you measured (in this case Start through Generation 4).

5. ☐ Use the graph below to plot the starting percent of the three genotypes listed in Table 4. Your instructor will show you how to join the dots using a BLUE pen.

6. ☐ On the same graph, plot the ending percent of the three genotypes listed in Table 4. Join the dots using a RED pen. Use the same method as in step 5.

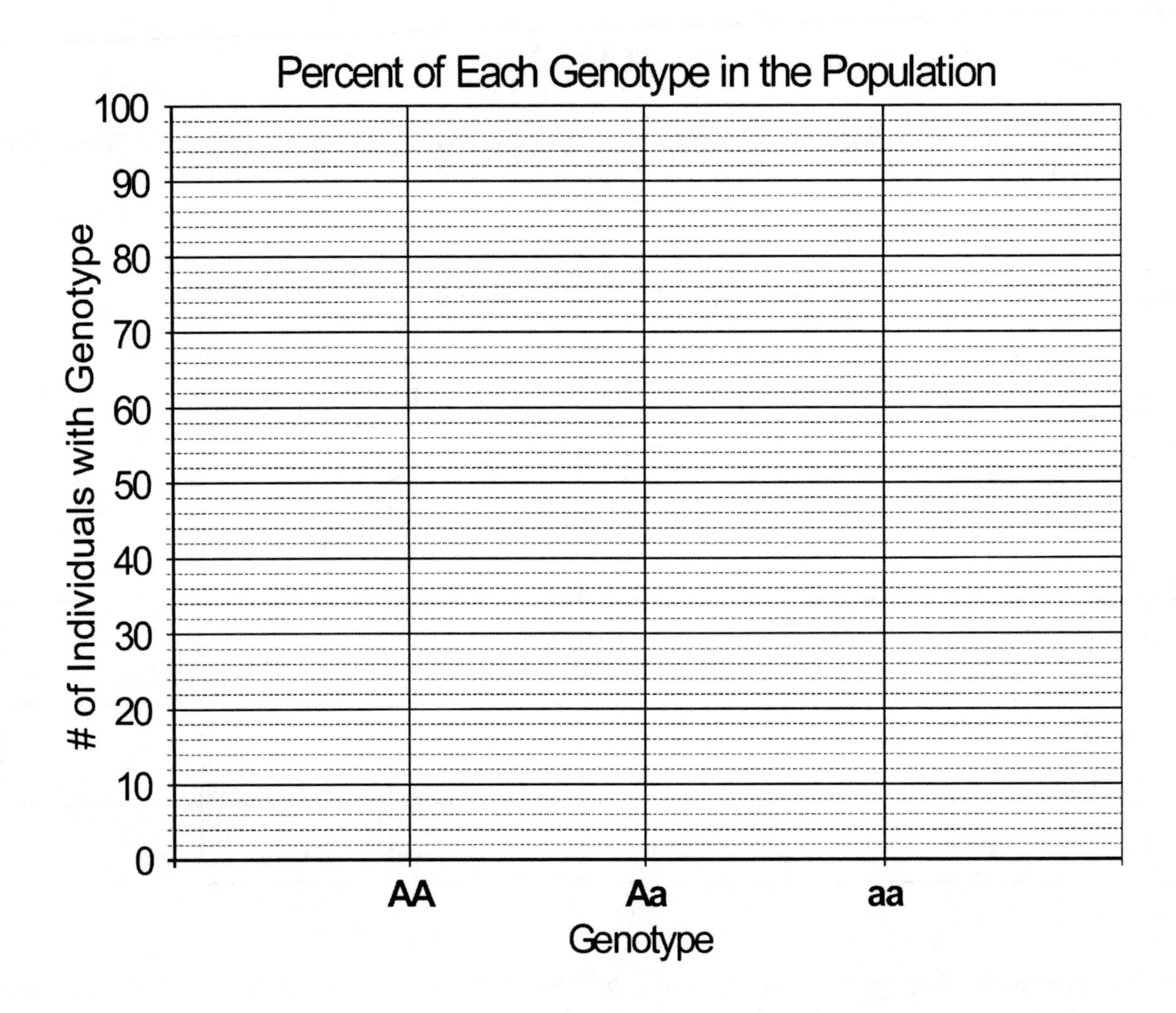

7. ☐ Look at the Modes of Selection diagram. Which Mode of Selection does your graph most look like?

Mode of Selection ______________________

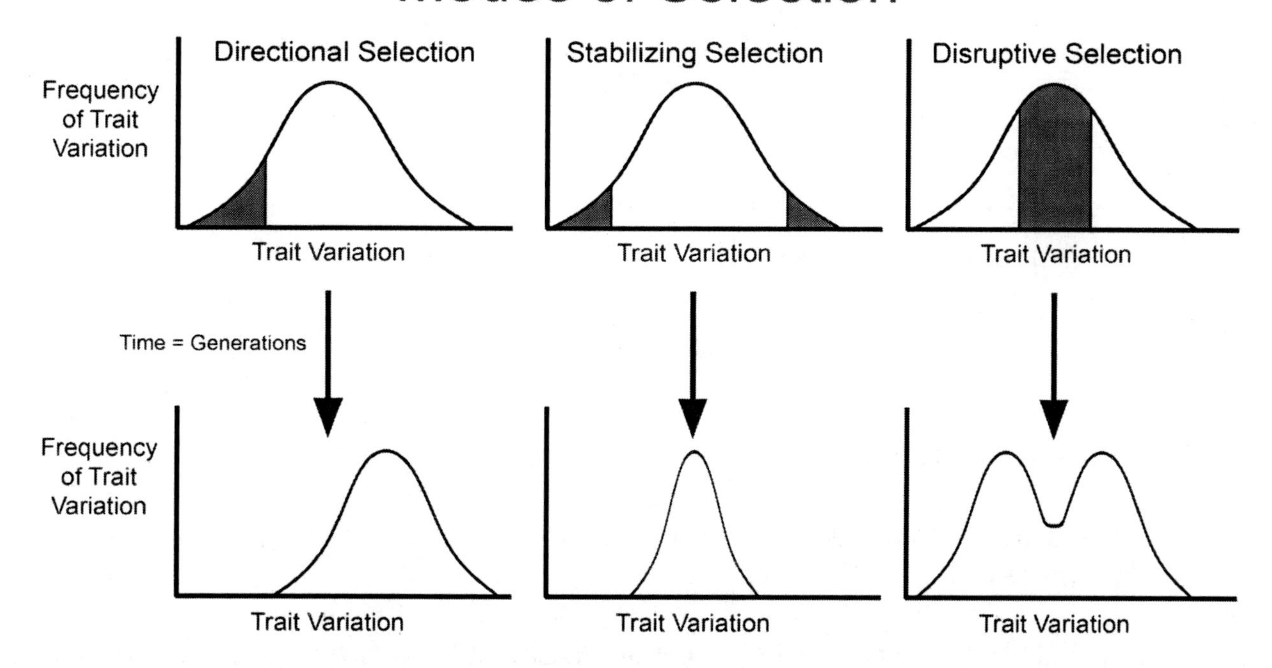

Directional selection results in a change in the mean value of the trait toward the form that has highest fitness.

Stabilizing selection results in the loss of the extreme forms of the trait; this means there is a decrease in genetic variation.

Disruptive selection results in an increase in both extremes and a loss of intermediate forms. A more complicated form of selection and if strong enough can lead to *speciation*, when groups of individuals with unlike extremes of a trait evolve to be so different that they can no longer reproduce with each other.

8. ☐ Now calculate the frequency of each *allele*, "A" and "a" (rather than the genotype like you did in above) in the starting and ending populations in Table 5. Start with the "A" allele for worm coloration. Using the table from step 4 calculate the number of "A" alleles in the starting population. Remember that each Black worm has two copies of this allele, Speckled has one, and Gray has none.

Table 5: Quantity of Each Allele in the Population Over Several Generations

Worm Color Alleles	Starting		Ending		Change from Start to End (%)
	#	%	#	%	
A					
a					
Total worms		100%		100%	

9. ☐ Next Calculate how many "a" alleles there are in your starting population.

10. ☐ Calculate the Ending allele frequencies in the population by using Table 4 "Ending" values.

11. ☐ Finally, calculate the percent change by, again, subtracting your Ending percent value from your Starting percent value for each worm allele color variation. Be sure to include the positive or negative sign of the calculated value.

    Why do we do calculate allele frequencies in the population? One way biologists measure evolutionary change is to observe changing allele or gene frequencies over time. So, we want to determine if some alleles for worm coloration are becoming more or less frequent in this population as time goes by.

Questions to Consider

What was the percentage of black, speckled and gray worms at the start?

In what way did the percentages change after a few generations?

What do you think would happen to the percentages of worm genotypes over many generations if the selection pressures remained the same as in the ending generation?

What are some of the factors that influence the number of surviving worms each generation?

What would happen if there was a sudden change in the environment that changes selection favoring speckled worms to black and gray worms? What mode of selection might then occur? List several types of environmental change that might cause this change in selection.

How does selecting for a phenotype affect the proportion of genotypes? Of allele frequencies?

Name: ________________________________ Instructor's Name: ______________

Normal Lab Day and Time: ________________

Natural World Lab Response Sheet

Lab 6: Reproducer – Natural and Sexual Selection

1. What was the main idea behind today's lab?

2. Why is having genetic diversity in a population important for natural selection to work?

3. Pretend that a population starts off with a bell curve distribution of female height. However, recently shorter females tend to have twice as many offspring as taller females. If this trend continues for several generations, draw a graph that would represent the new distribution of female height in this population. Be sure to label your graph axis.

4. What mode of selection does the graph you made most look like?

5. Rank today's lab from 0 (poor) to 10 (excellent). <u>WHY</u> did you choose this rating?

LAB 7: PLANT REPRODUCTION – GETTING INTO THE FUTURE

Sexual reproduction is a way for plants, and most other organisms, to create and disperse their offspring. These offspring share genes with their parents and carry on those genes through time and space. Sexual reproduction is also a way to make novel combinations of genes, a process that is important for maintaining genetic diversity. Through reproduction plants and their genes get spread around. Getting their offspring spread out over a large area is advantageous in that it makes it more likely that they'll end up in a place where the offspring will survive and pass on the parents' genes.

But reproduction is also risky for plants. In the process of getting pollen (which contain sperm) to the female structures of another flower of the same species nearly all pollen grains are lost, and most pollen falls near the paternal plant. The same is true of seeds – nearly all are lost and most fall near the maternal plant. Plants have developed a number of strategies to avoid squandering their genetic investment in pollen and seeds.

Plants cannot themselves move, at least not very quickly, so they take advantage of things that can. Many plants use the wind to carry pollen or seeds to their targets. Casting your pollen or seed to the wind is not very targeted, and you can imagine that plants that use the wind for pollination or seed dispersal have to make a lot of pollen or seeds. Many of us have personal experience with the fact that a lot of wind-borne pollen doesn't reach its target – we get hay fever.

Other plants increase the chance that their pollen or seeds will get to the right place by making use of animals as couriers or *vectors*. Plants "reward" their animal vectors for this service by offering them sugary nectar and/or high-protein pollen for pollination or juicy and nutritious fruit for seed dispersal. Plants "advertise" these rewards by producing showy flowers, delicious odors, and colorful fruits. The relationship between plants and their pollinators or seed dispersal vectors can be very specific. Some flowers, for example, look and smell remarkably like the female bees that pollinate them. A butterfly's tongue is the same length as the tubular flower it pollinates. In other cases, plants act as hitchhikers on animals by attaching their seeds or fruits to fur or feathers.

Even if sexual reproduction is successful and a seed does manage to get into a suitable habitat and germinate, it still faces a lot of challenges: the seedling has to compete with other seedlings and preexisting plants for light, water, space, and nutrients. The mother plant is an especially strong competitor as it is a member of the same species and has the same ecological requirements (a good reason for the seedling to get as far from its parent as it can). The seedling will also face disease organisms, herbivores, and the variability of weather and climate.

In this lab, you will explore some of the strategies plants use to get their seeds moved around and how plants get pollinated.

Flower Parts

Flowers are complicated reproductive structures that vary from one species to the next. For example, some flowers have few stamens and others have many. Some have one pistil, others have several or many. In some, the flowers have the petals fused into a single structure while others have petals that are separate from one another. The ovary – the part of the pistil that will develop into the seed-containing fruit – can be located either above or below the other flower parts. Many other differences in number, position, and fusion are possible. Below is an illustration of several flower shapes. On the following page the major floral parts are illustrated and identified. Your instructor will point out several points pertaining to the position of the ovary or ovaries.

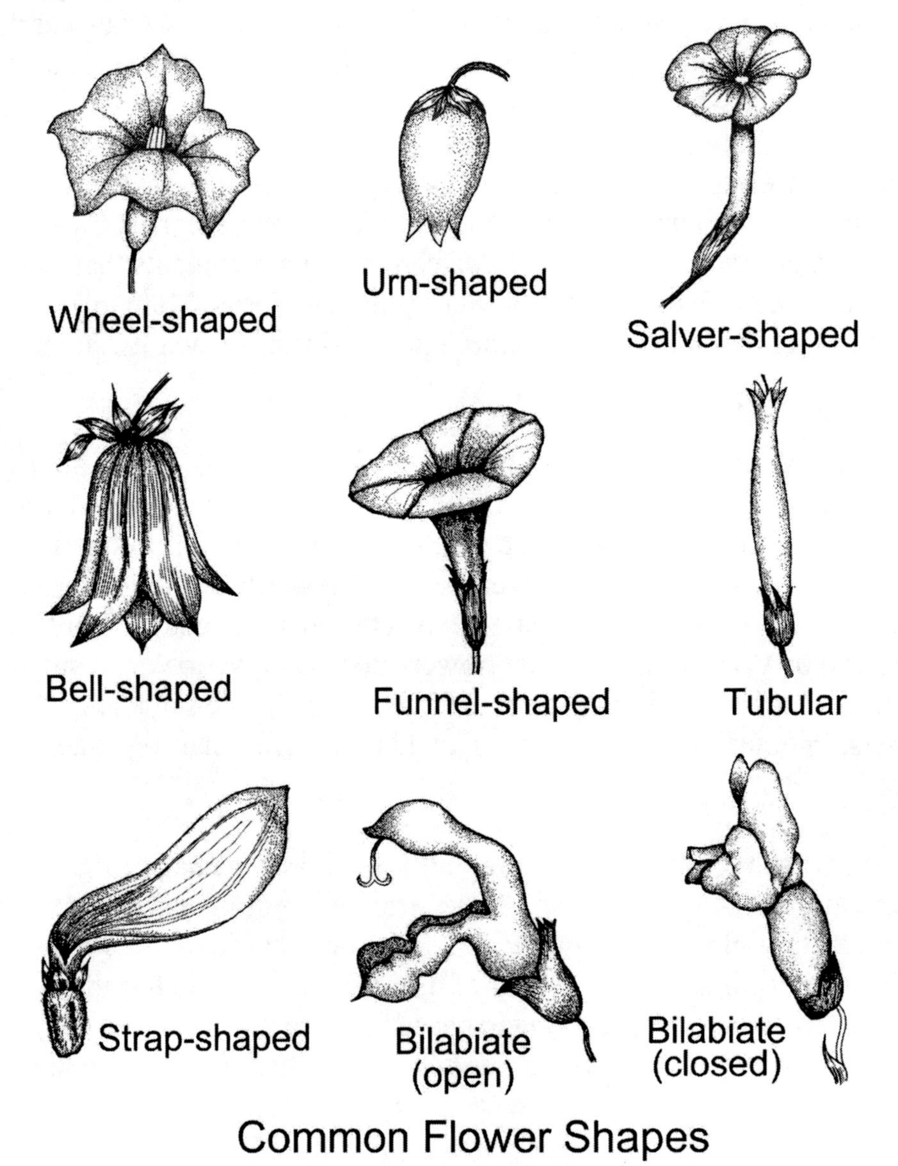

Common Flower Shapes

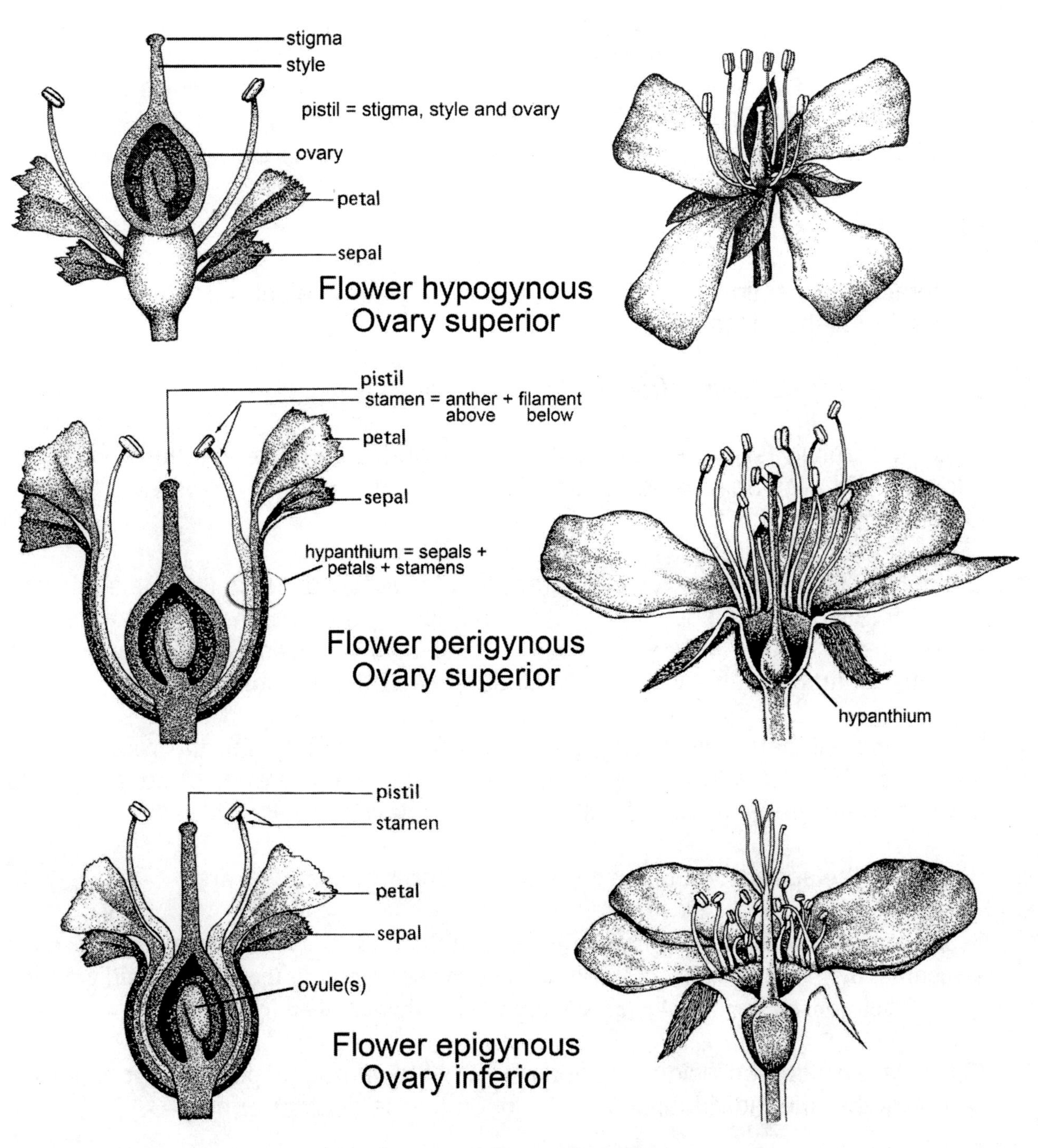

Flower Morphology

Activity 1: Flower Morphology

Flowers come in many different shapes, colors, and sizes, but all have parts involved with the reproduction of the plant. We are going to dissect flowers to learn about their parts and investigate the similarities and differences in their reproductive anatomy. Note that in some species the male and female reproductive parts are in different flowers (corn) or even on different plants (cottonwood trees). Many flowers have markings – like lines or spots – that guide pollinators to the "business end" of the flower. These markings are called nectar guides.

Procedures

1. ☐ Obtain a dissection tray, scissors, razor blade or scalpel, forceps, dissecting needle, and a hand lens.

2. ☐ Pick one of the flowers from those provided.

3. ☐ Look carefully at the shape of the flower and determine its shape from the "Common Flower Shapes" diagram. Fill this out in Table 1: Flower Morphology Observations.

4. ☐ Use the "Flower Morphology" diagram and to identify the sepals, petals, pistil(s) and stamens on your specimen without damaging the flower. Note whether the stamens rise above or are below the level of the petals. If you cannot see everything, that's okay – we'll take a closer look at the structures later.

5. ☐ Now we will begin dissecting the flower to more clearly see the reproductive structures. Use the "Flower Morphology" diagram as a reference. Also, as you go through these steps, be sure to fill out the information in Table 1.

6. ☐ Identify the sepals on the flower and count how many there are.

   Sepals are the outermost structures at the base of the flower. Typically they are green but sometimes they are colored and petal-like. They can be as large as the petals or withered and small. Their job, an important one, is to protect the flower when it is in bud.

7. ☐ Petals are attached inside the sepals. Count the number of petals on your flower. Note whether the individual petals are fused to one another or not.

   Petals are sometimes not present in some species. In these cases the sepals are often colored and play the part of the petals – being highly attractive to pollinators. Sepals and petals are not primary reproductive structures, but do play an important part in pollination as we will see in Activity 3.

8. ☐ Gently remove the sepals and petals of the flower to reveal the sex organs.

Flowers are usually *hermaphroditic*, meaning they contain both male and female sex organs. In this case the male organs are the *stamens* and the female organs are the *pistil*.

9. ☐ Locate the stamens and identify their *anthers* which are the small sacs that produce the sperm-containing pollen.

The thousands of pollen grains produced in the anther will be dispersed by one of several modes of dispersal, such as wind, water, and insect pollinators (see Activity 2). Each pollen grain contains two sperm cells (only one in gymnosperms). The stalk that holds and supports the anther, as well as exposes it to the plant's specific pollen dispersal mode, is the *filament*.

10. ☐ Carefully snip off the filaments to expose the pistil. The pistil consists of the stigma (top part that receives pollen), a thin style (which sometimes is very short), and ovary (the base which contains the ovules) as indicated on the "Flower Morphology" diagram.

In order to catch and retain pollen grains the stigma typically secretes a sticky substance and/or is feathery. The pollen grains germinate on the stigma and grow a tube that travels down through the style and ultimately into the ovary. The ovary contains ovules, each with an egg that the sperm traveling down the pollen tube can then fertilize.

11. ☐ Looking at the "Flower Morphology" figure, what is the position of the ovary in the flower. Is it superior (above the sepal and petals) or inferior (below the sepal and petals)?

12. ☐ Use a razor blade or scalpel to cleanly slice the ovary LENGTHWISE. Expose the ovules by gently opening the ovary with a teasing needle. You may need to use a hand lens or dissecting scope to see the ovules. Count the number of ovules if they are visible. If there are more than 20, simply say "many."

After an egg and sperm unite, the female reproductive parts of the plant drastically change. The petals and stamens fall off and the ovary swells to form a "fruit." *What do the fertilized ovules become in the fruit?*

13. ☐ Go to the section below entitled "Flower Power." Read the introduction and review Table 2. Using this information, which mode of pollination does your flower(s) most likely use? Why did you choose this mode of pollination?

Table 1: Flower Morphology Observations

Characteristic	Flower Name:________________
Flower Shape	
Sepal Description (color and number)	
Petal Color	
# of Petals (count lobes if fused)	
Nectar Guides (present or absent)	
Stamens/Filaments: (extended above or below petals?)	
# of Stamens	
Stamen height compared to pistil? (above, same, or below)	
Hypanthium (present or absent)	
# of Pistils	
# of Stigma	
Ovary Position (superior or inferior)	
# of Ovules (if possible to count)	
Type of Pollinator (see Table 2)	

Fruit and Seed Dispersal

If fertilization is successful, the sperm and egg fuse to form a zygote which grows into the embryo in the seed. Seeds will develop within the plant's ovary from the fertilized ovules. The mature seed-containing ovary is called a *fruit*. If a plant organ contains seeds it *is* a fruit. So, the age-old question is answered: yes, tomatoes are fruits (as are okra, squashes, peppers, green beans, zucchini, and many other edibles that get called "vegetables"). Seeds and fruits are "designed" to get dispersed so that the parental plants can spread their genes. For example, some seeds and some fruits have feathery plumes that enable them to get around on the breeze. In others the wall of the ovary develops in a way that increases the chance that the seeds will be dispersed by animals eating them. Fruits are often sweet, fleshy, and juicy. Many seeds are dispersed after passing through the digestive tracts of animals that found their enclosing fruits to be a tasty, nutritional or water-providing addition to their diet. In fact, many seeds *must* pass through an animal before they can germinate! Other fruits don't appeal to an animal's sense of taste as they are dry, hard, or fibrous. These are more often dispersed by the wind (like dandelions) or by attaching themselves to feathers or fur (cockleburs). Yet other fruits use pressure to ballistically eject their seeds under pressure (mistletoe), resemble salt shakers (poppies), or use water (mangroves) to move around. Oak trees are dispersed by squirrels burying their seed-containing fruits (acorns) in the fall. Squirrels often forget where they buried these fruits and, hence, oak trees "get into the future."

Activity 2: Flower Power

Plants have a problem with sex – they usually need help to get the sperm-containing pollen to the receptive part (the stigma) of the female pistil. If they don't use wind, water, or gravity to do this, they must rely on animal vectors. Not surprisingly, many plants have adaptations that help to attract specific pollinators to their flowers. The relationship between flowering plants and pollinators involves mutualism – the pollinator receives a reward (nectar, some of the pollen, or occasionally a usable scent) while the plant benefits by the pollinator spreading its pollen. Some pollinators, as already mentioned, use the flower's scent as a reward. These scents are typically then used as a sexual attractant. One wonders if any non-pollinating animals use the scents of flowers as sexual attractants. Hmm . . . .

On the next page is a summary of some of the characteristics of major plant pollinators and of the traits of flowers that attract them. Bear in mind that some flowers use more than one pollinator – these "generalists" will take advantage of any willing pollinator passing by. Also, some flowers have been highly manipulated by us to exhibit traits like color, flower size, fragrance, or number of petals, that we find beautiful or interesting. These modified traits may not be in the plant's best interest in the wild. The flowers of any particular plant species could posses several of the characteristics listed in the table below.

Table 2: Flower Pollination Characteristics

Pollinator	Pollinator and Flower Characteristics
Bee	Bees have excellent eyesight, require high-energy rewards, and can see ultraviolet light as a distinct color; they can recognize and remember colors, odors, and shapes. Bee flowers are typically fragrant, brightly colored, usually blue, purple or yellow, often have nectar guides and landing platforms, and are often irregular in shape.
Bat	Bats have good vision, but are active at night and, therefore, are not able to see color. They have a good sense of smell and prefer fruit-like, fermenting, or musty odors. Bats perch by hanging upside down and lap nectar or gather pollen, which is high in protein. Bat flowers are typically white or light colored, often hang down or, when facing up, are bowl shaped. Flowers are strong and have a bat-preferred smell.
Butterfly	Butterflies have good senses of sight and smell and can perceive red as a distinct color. They drink nectar from a long proboscis; flowers are orange, yellow, blue and even red, are tubular (why?), and have a strong, sweet smell.
Hummingbird	Hummingbirds have a poor sense of smell but highly developed vision. When they are drinking nectar, they prefer to tip their heads back while hovering. Bird flowers are brightly colored and are often red or yellow. They are also medium sized to large, tubular, hanging down, and have no smell. Pollen is placed so that it can be transferred onto the bird's throat or head. Nectar is low in sugar (why?)
Beetle	Beetles have a good sense of smell but poor eyesight. They prefer fruity, spicy, or fermentation-like smells. Flowers are round in outline, white or dull colored, and often with a flat, broad, landing platform of petals. Reproductive parts of the flower are exposed.
Fly	Flies have good vision and sense of smell. Flowers may or may not be brightly colored, but they have a strong odor that often smells rotten, sweaty, or of feces. Think about where flies like to congregate.
Moth	Moths have good senses of sight and smell, are typically active at night, and prefer strong sweet smells. Flowers are white (why?), have a tubular shape, and have a strong, heavy, sweet odor.
Wind	Petals and sepals of these plants are very small or lacking all together, and the flowers have no odor. Stigmas are large and feathery. Copious amounts of pollen are produced from protruding stamens.
Water	Flowers are flush with the water's surface. Pollen is released on the water and floats. Stigmas are flush with the water. Flowers are small and scentless. A very rare form of pollination.

Procedure

□ You are to investigate various adaptations for pollen dispersal in several plants available to you either in the lab or in the greenhouse. Use the information in the above table and in the introduction to today's lab to determine a *probable* mode of pollen or seed dispersal of the plants you are examining. Fill in the table below. Be sure to justify why you chose that particular pollinator in the "Supporting Evidence" column.

	Specimen (name)	Specific Pollinator	Supporting Evidence (be specific)
1			
2			
3			
4			
5			
6			
7			
8			
9			
10			
11			
12			
13			
14			
15			

Activity 3: Fruit and Seed Examination

Procedures:

1. ☐ In small groups or individually, examine the fruits and/or seeds provided. You may need to use a dissecting microscope to examine some of the samples.

2. ☐ For each fruit or seed type, describe its mode of dispersal (kind of animal or process), and list the adaptations for that type of dispersal in Table 3 below. Make use of all of your senses!

Table 3: Fruit and Seed Observations

Fruit or Seed Type	Dispersal Mechanism	Adaptations

Questions to Consider

What are some advantages to "getting around" or dispersing?

What are some risks or disadvantages of dispersing?

What are some common adaptations for wind-dispersal?

How do plant seeds or fruits hitch rides on animals?

List some examples of how flowers are adapted to their pollinators and vice versa.

Name: ______________________________ Instructor's Name: ____________

Normal Lab Day and Time: ______________

Natural World Lab Response Sheet

Lab 7: Plant Reproduction Adaptations

1. What was the main idea behind today's lab?

2. Look at the fur piece on your table. Do any of the seeds have another mode of seed dispersal other than getting attached to fur? Compare the seeds that are fur only dispersal with those that could also be dispersed by wind. What are the differences between them and how do these aid with their mode of dispersal? Be sure to draw one example of each – labeling the parts you discussed.

3. What *specific* part of a flower contains the pollen that gets "rubbed" onto the heads of hummingbirds?

4. Rank today's lab from 0 (poor) to 10 (excellent). WHY did you choose this rating?

LAB 8: PLANT NUTRITION AND SYMBIOSIS

Perhaps you have a houseplant that is looking a little sickly. You can purchase some soluble plant food, mix it in a watering can, and water your plant with it. If the problem was a lack of soil nutrients, then you can expect to see good results within a few weeks.

Did you "feed" your plant with the "plant food?" Not really. Plants are organisms that manufacture their own food – the energy source glucose – through photosynthesis. It is true, however, that plants have requirements beyond the carbon dioxide (CO_2), water, and light they need to produce glucose by the process of photosynthesis. Soil provides plants with many nutrients. The nutrients plants need in the largest amounts are nitrogen, phosphorus, and potassium. These are the N, P, and K you see listed on a box of "plant food" or fertilizer. This lab will reveal the effects different concentrations of nitrogen and phosphorus have on plants.

Plants use nitrogen to build chlorophyll molecules, nucleic acids (DNA and RNA), and proteins, like the enzymes involved in photosynthesis. Lack of nitrogen has some of the most obvious effects on plants. Can you guess what at least one of those effects might be?

Although nitrogen makes up almost 80% of the atmosphere, it is not available to plants in its abundant molecular form of N_2. The nitrogen that plants can use is in the form of ammonia (NH_3), nitrite (NO_2^-), or nitrate (NO_3^-). Beneficial bacteria called *Rhizobium* in the roots of legumes (plants in the bean family) *fix* nitrogen into usable forms in exchange for carbohydrates from photosynthesis. Lightning, volcanoes, and free-living photosynthetic bacteria are other natural sources of fixed nitrogen. These days, the amount of nitrogen fixed artificially, through the manufacturing of fertilizers, equals the amount fixed by natural processes.

Phosphorus is an important constituent of cell membranes, it is the "P" of ATP (the energy currency for cellular processes), and forms the backbone of DNA and RNA. What do you think happens to the growth of a plant that lacks phosphorus? Since growth means making more cells, and phosphorus is an important part of several cellular components, then how well can a plant grow with too little phosphorus?

The phosphorus in plants ultimately comes from the slow decomposition of rocks. In soil phosphate clings to clay particles. Since phosphate doesn't move much in the soil, plants have to go to great lengths (literally) to get it. Plants can do this in one of two ways: they can grow long, finely-divided root systems, or they can enter into a *mutualism* (mutually beneficial symbiotic relationship) with fungi that help them to acquire phosphate. Fungi that have a mutualistic relationship with plant roots are called *mycorrhizae* and occur in at least 90% of plant species.

In Activity 1 we will set up an experiment to test how tomato seedlings respond to nitrogen and phosphorus. We will include treatments lacking nitrogen (N) or phosphorus (P) as well as treatments with twice the normal N or P to see if it is possible to get too much of a good thing. In Activity 2 we will observe and measure clover plants to see the effect of *Rhizobium*. Activity 3 examines rhizobial root nodules and Activity 4 involves microscopic observation of mycorrhizal fungi.

Activity 1: Tomato Seedlings & Nutrients

Procedure 1 (set up for future observation):

1. ☐ Work in groups of 3-4 students.

2. ☐ Label six plastic cups with your lab time, group name, and each of the following treatments:

    - **H_2O** pure water
    - **CNS** complete nutrient solution
    - **N-** solution with proper amount of phosphorous but *without nitrogen*
    - **N+** solution with proper amount of phosphorous but *twice the normal nitrogen*
    - **P-** solution with proper amount of nitrogen but *without phosphorus*
    - **P+** solution with proper amount of nitrogen but *twice the normal phosphorus*

3. ☐ Fill each cup 3/4 full with the appropriate nutrient solution (or water) and cover with a square of aluminum foil.

4. ☐ Punch a small hole in the foil on each cup.

5. ☐ Pick out six tomato seedlings and gently shake the soil from their roots. Assign each seedling to one of the above treatments.

6. ☐ Keep each seedling in a moistened, folded paper towel while doing the following measurements: weigh each seedling (without the towel) on a balance set to grams (g) and measure the length of each plant in cm. (Measure from the bend in the root/stem to the tip, or growing point, of the top most leaf. *See the following figure*.

7. ☐ For each plant, record the plant weight, length, leaf color, and the color of the *cotyledons* in the table on the next page for "Initial Observations."

The cotyledons are the first leaves to appear as a seed germinates. They are sometimes referred to as seed leaves and usually look different than the true leaves. True leaves are all of the leaves other than the cotyledons. True leaves tend to look like the type of leaves you would expect to see on a plant of that species. See the illustration below.

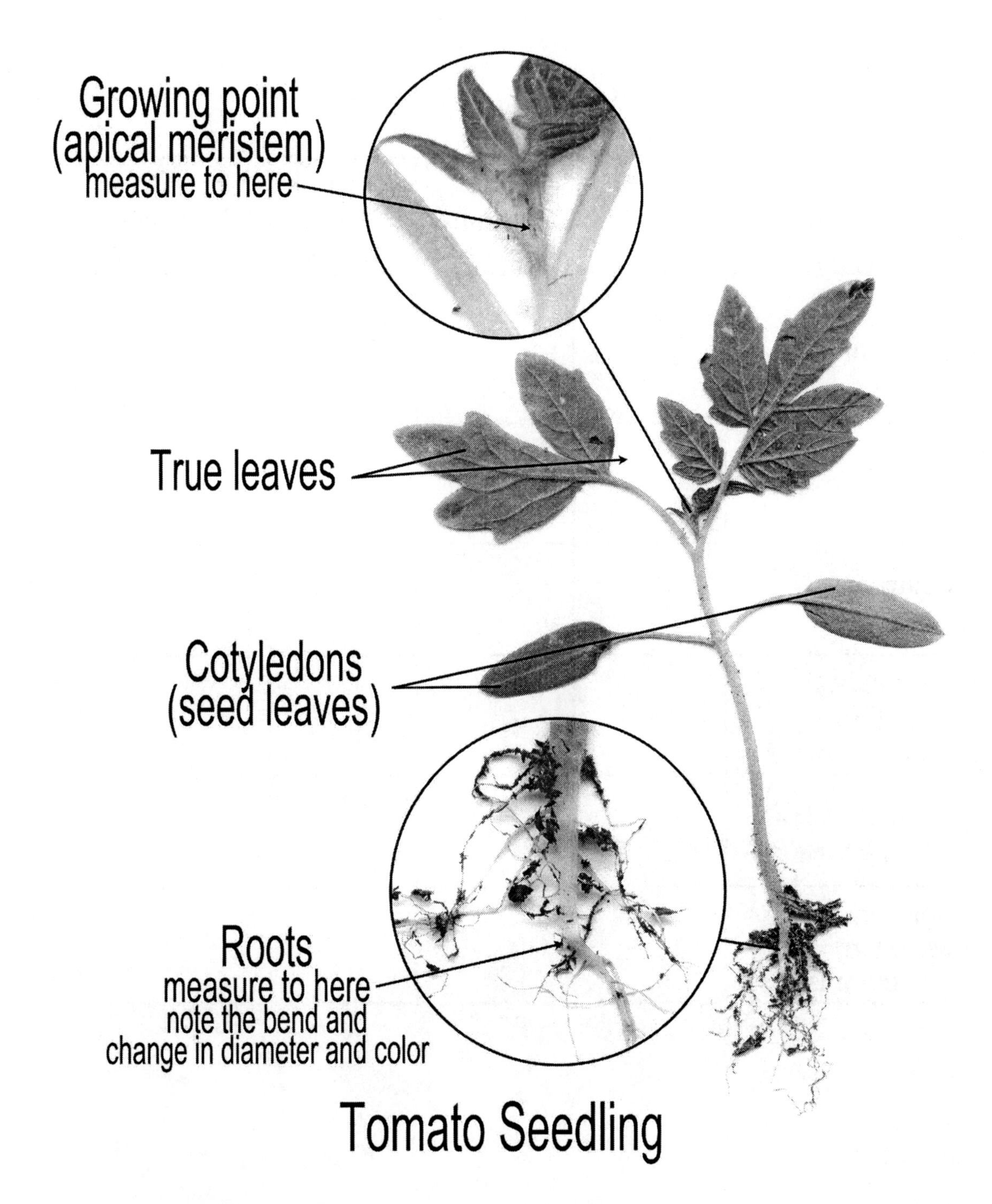

8. ☐ Carefully place each seedling through the hole in the foil on the appropriately labeled cup so that the roots are in the nutrient solution.

9. ☐ Place the set of six plants in a tray under the grow-lights. We will make a second set of observations of the tomato plants after 2–3 weeks.

Initial Observations

Treatment	Plant Weight (g)	Shoot Length (cm)	Color of True Leaves	Cotyledons: Presence and Color
Pure water				
Complete Nutrient Solution (CNS)				
Solution without Nitrogen (N-)				
Solution with twice the Nitrogen (N+)				
Solution without Phosphorus (P-)				
Solution with twice the Phosphorus (P+)				

Final Observations (After 2–3 weeks)

Treatment	Plant Weight (g)	% Change*	Shoot Length (cm)	% Change*	Color of Leaves, Top and Bottom	Cotyledons: Presence and Color
Pure water						
Complete Nutrient Solution (CNS)						
Solution without Nitrogen (N-)						
Solution with twice the Nitrogen (N+)						
Solution without Phosphorus (P-)						
Solution with twice the Phosphorus (P+)						

***To calculate the percent change:**

$$\left(\frac{\text{Final Value} - \text{Initial Value}}{\text{Initial Value}}\right) \times 100 = \textit{percent change}$$

Questions to Consider

Which plants changed the most during the course of the experiment? Were the changes you observed for "better" or "worse?"

Is it possible for a tomato seedling to have too much of an essential nutrient?

What color were plants growing in a solution without N? How is their color related to the chemical composition of a particular molecule in plant leaves?

When plants were grown in a solution with twice the normal amount of N or P, what happened to the color of the solution by the end of the experiment? Was there anything growing in it?

If a farmer applies more nitrogen fertilizer to a field than the crops can use for growth, what happens to the excess?

Which do you think would grow better: a plant in "complete nutrient solution" or a plant in good, fertile soil? Explain your choice.

Activity 2: Effects of Nitrogen Sources on Clover Plants

Procedure:

1. ☐ Each group of students should obtain three pots of clover, one from each of the three treatments: A, B, and C. Each pot should have the same number of plants in it. Your task is to identify which pot received which of the following treatments:

   R+ F-: plants with *Rhizobium* bacteria, no additional nitrogen fertilizer added
   R- F-: plants without *Rhizobium*, no additional nitrogen fertilizer added
   R- F+: plants without *Rhizobium*, nitrogen fertilizer added

2. ☐ Observe the general color and condition of the plants and record them in the table on the next page.

3. ☐ On a sheet of newspaper, dump out the contents of each pot. **Keep the treatments separate and make sure you know which is which!**

4. ☐ Gently shake the roots to remove most of the soil and then rinse off the rest over the sink.

5. ☐ Look for the presence or absence of small, tumor-like ***nodules*** on the roots. Record the results in the table below.

   Clover plants form root nodules in response to inoculation with *Rhizobium* bacteria. In the nodules the plant provides the bacteria with high-energy carbohydrates and the *Rhizobium* fixes nitrogen which is available to the plant.

6. ☐ Blot the now soil free roots dry and weigh all the plants from each pot (not as individual plants but as a whole pot). Make sure that the balance is set to grams (g). Record your results in the table on the next page.

Pot	General Appearance and Color of Plants	Nodules Present ?	Plant Weight (g) per Pot	Treatment (R+F-, R-F+, or R-F-)	Did it get nitrogen? If so, from where?
A					
B					
C					

7. ☐ In the final column of the table, assign the correct treatment designation to each of the pots.

Activity 3: Examination of Root Nodules

Procedure:

1. ☐ Pick off two nodules. Be sure they are washed and dried.

2. ☐ Cut one nodule in half with a razor blade and leave it exposed to the air for a few minutes. Look for the development of a pinkish color due to the presence of leghemoglobin.

   Leghemoglobin, like the hemoglobin in your blood, has a strong affinity for oxygen. *Rhizobium* bacteria require a low oxygen environment. In the root nodule, leghemoglobin binds up oxygen so that the bacteria get the low-oxygen environment they need. This is an amazing symbiotic interaction! The "heme" of hemoglobin comes from the bacterium and the "globin" comes from the plant.

3. ☐ Take another nodule and place it sandwich-style between two microscope slides. Smash it between the slides. Then, slide one of the glass slides over the other to smear out the smashed nodule.

4. ☐ Let the smear dry.

5. ☐ Choosing the slide with the thinnest amount of nodule "goo" (no chunks please), heat-fix the smear by passing the slide, smear side up, twice through a Bunsen burner flame.

6. ☐ Flood the smear with methylene blue dye and leave the dye on for 30 seconds.

   Methylene blue stains the *Rhizobium* bacteria so that they are visible.

7. ☐ Rinse the slide and gently ***blot*** dry.

8. ☐ Focus under low power, then change objectives to high power. Ask your instructor for help using the oil immersion lens, if your microscope has one.

9. ☐ Look for clusters of oval bacteria inside infected cells. Draw some in the space below.

Questions to Consider

Which was more effective at promoting healthy growth of clover plants: *Rhizobium* inoculation or nitrogen fertilization? Or was there no difference?

What do the bacteria provide to the plant? How does the plant "pay" for this service?

Imagine two clover plants: one in a nutrient-rich soil and the other in a nutrient-poor soil. If you innoculate (infect) both of their soils with *Rhizobium*, in which soil would you expect a greater effect? Explain.

What was the point of the treatment that involved plants without *Rhizobium* but with additional nitrogen fertilizer?

Would soybean plants in a heavily fertilized field benefit from inoculation with *Rhizobium*? Do you think you would find many nodules on roots in this situation?

Activity 4: Mycorrhizal Fungi

Procedure 4:

1. ☐ Obtain a prepared microscope slide of roots with mycorrhizal fungi.

2. ☐ Observe the specimen at 100× total magnification.

   Root cells will have light-brown stained walls and appear mostly "hollow." Mycorrhizal hyphae will look like dark blue to black twisted threads. You may be able to find vesicles, fungal storage structures, which look like bluish-gray oval balloons inside plant cells. Arbuscules are branched fungal structures inside plant cells. They are thought to function in nutrient exchange between the fungus and plant.

3. ☐ Draw a portion of what you see below, labeling root cells, mycorrhizal hyphae, vesicles, and arbuscules (if found).

Questions to Consider

What does the term *mutualism* mean? Give two examples.

If you looked, would you find more mycorrhizae in nutrient-poor or nutrient-rich soil?

Fungicides are used to kill disease-causing fungi on crop plants. What effect might this have on the plants' normal relationship with mycorrhizal fungi that are in the soil? Explain how the fungicide might affect the health of plant?

Name: ______________________________ Instructor's Name: ____________

Normal Lab Day and Time: _______________

Natural World Lab Response Sheet

Lab 8: Plant Nutrition

1. What was the main idea behind today's lab?

2. Why do many plants have *mycorrhizae* rather than just growing a more developed root system? (Reading the bottom half of page 1 of the lab will help you answer this question.)

3. Which do you think would do better...a plant given a complete nutrient solution or a plant that had been inoculated with *Rhizobium*? Why? *Recall the clover results in Activity 2 to help you think through the answer to these questions.*

4. Rank today's lab from 0 (poor) to 10 (excellent). <u>WHY</u> did you choose this rating?

Lab 9: Soil – The World Beneath Your Feet

Soil is a very complex "body" of substances. It consists of minerals, air, water, living organisms (lots of them), and decomposed organic matter. It has a history. It "grows," changes, and can be "killed." It is virtually a living organism. Most of us, however, don't fully appreciate the importance of soil – it's just "dirt" after all. Some quick reflection will convince you that all of the terrestrial vegetation in our world ultimately depends on the soil. Soils nurture and support *all* land plants and the animals that eat them. These are the primary sources of our food and shelter (think wood). Soil, then, is likely our most precious natural resource. In this investigation you will examine some of the characteristics and living inhabitants of this amazing stuff we call soil, both *inorganic* (chemical) and *organic* (living or derived from living organisms).

Part I: Physical Properties of Soil

Activity 1: Texture

Soil texture is a function of the combined amount of sand, silt, and clay present. Course-textured sandy soils hold little water, let air circulate freely, and have few nutrients. Floury silty soils hold about the right amount of air and water but typically do not contain many nutrients. Fine-textured clayey soils let little air in but hold onto water so tightly that much of it is not available for plants to use. Clayey soils may have a lot of nutrients, but because of poor air and water relations, plants do not do well in soils high in clay. Soils with a balance of sand, silt, and clay are called *loams* and have the best balance of air, water, and nutrient characteristics.

Your instructor will go over some of the properties of pure sand, silt, and clay. Take some brief notes for use later in this investigation.

Sand:

Silt:

Clay:

In the ***Soil Pyramid*** below, you can see how the proportions of sand, silt, and clay are used to determine a particular soil's overall type of texture. Your instructor will show you how to read the Soil Pyramid.

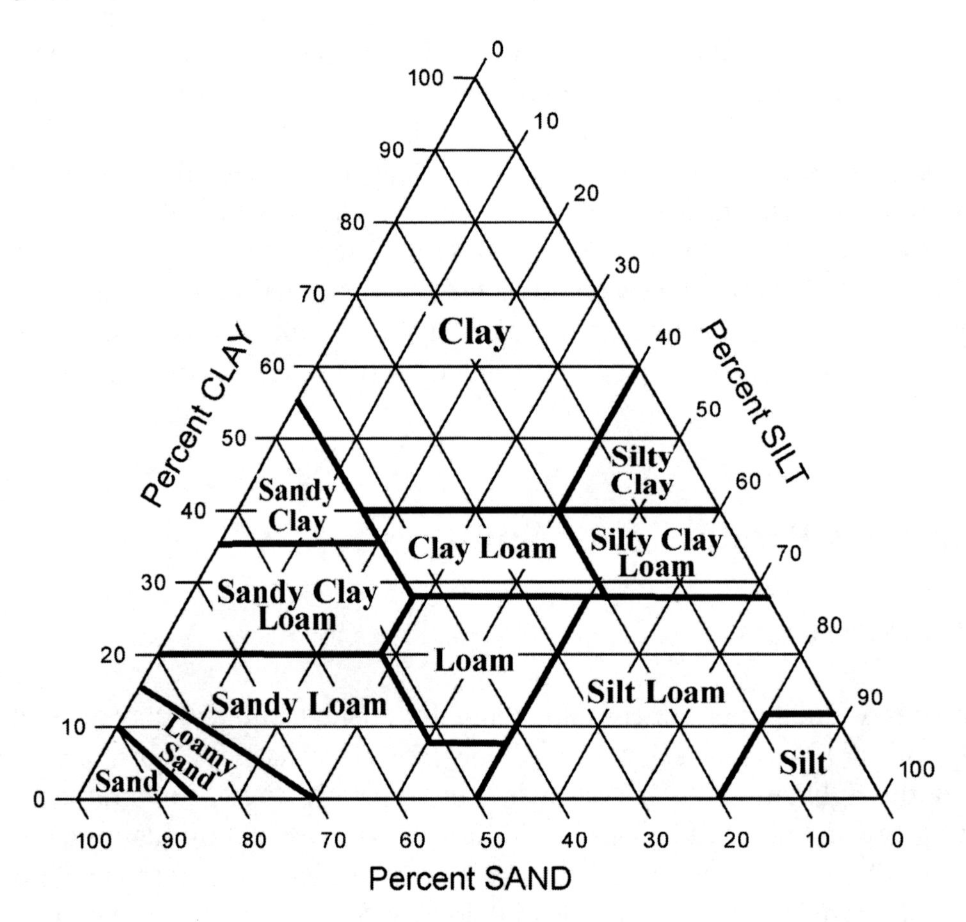

For the first part of this Activity, you will determine the texture of two or three soil samples. Believe it or not, you will be estimating the proportion of sand, silt, and clay in your samples by getting your hands dirty! This really is how county extension agents typically get the job done.

Procedure:

(Modified from the NASA website ltpwww.gsfc.nasa.gov/globe/index.htm)

Step 1. Place an amount of soil about the size of an egg in your hand. Use a mist bottle to moisten (not soak) the soil. Let the water soak in and then, as if it were dough, work the soil until the moisture is distributed throughout. Once the soil is moist, try to form a ball by squeezing it in your hand.

(a) If the soil does not form a ball at all, it is ☐ **SAND.** (check if it applies)

(b) If the soil forms a ball (even if very fragile), go to Step 2.

Step 2. Try to form a "ribbon" with the moistened soil. With the wetted ball of soil in your hand, begin squeezing it with your thumb and forefinger. Make the longest ribbon you possibly can while being as gentle as possible!

If the soil forms

(a) a long ribbon (more than 5 cm long), it is a kind of **CLAY**. Go to Step 3.

(b) a medium ribbon (2–5 cm long), it is a kind of **CLAY LOAM**. Go to Step 3.

© a short ribbon (less than 2 cm long), it is a kind of **LOAM**. Go to Step 3.

(d) no ribbon, go to Step 4.

Step 3. Place a small ball of soil about the size of a small marble in the palm of your hand. Add extra water to it so it's a little runny. Rub the paste in the palm of one hand and feel for grit with the thumb of your other hand.

If the soil:

(a) feels very gritty (a large amount of sandy particles), add the word **SANDY** before the name you found above

☐ **SANDY CLAY**
☐ **SANDY CLAY LOAM** (check the appropriate box)
☐ **SANDY LOAM**

(b) feels only a little gritty (some but not a lot of sandy particles), use just the name you found above

☐ **CLAY**
☐ **CLAY LOAM** (check the appropriate box)
☐ **LOAM**

© feels very smooth, with <u>no</u> gritty feeling at all, add the word **SILT** or **SILTY** before the name you found above

☐ **SILTY CLAY**
☐ **SILTY CLAY LOAM** (check the appropriate box)
☐ **SILT LOAM**

Step 4. Take about 1/5 of your ball of soil and add extra water to it so it's a little runny. Rub the paste in the palm of one hand and feel for grit with your thumb or finger tips.

If the soil:

(a) feels gritty, it is a ☐ **LOAMY SAND**

(b) feels soft and smooth with no gritty feeling, it is a ☐ **SILT**

Activity 2: Field capacity

Field Capacity refers to how much water a soil can hold after gravity has removed the "excess." The graph below shows the average field capacity for the various soil types. Notice that sandy soils hold the least amount of water and clayey soils hold the most.

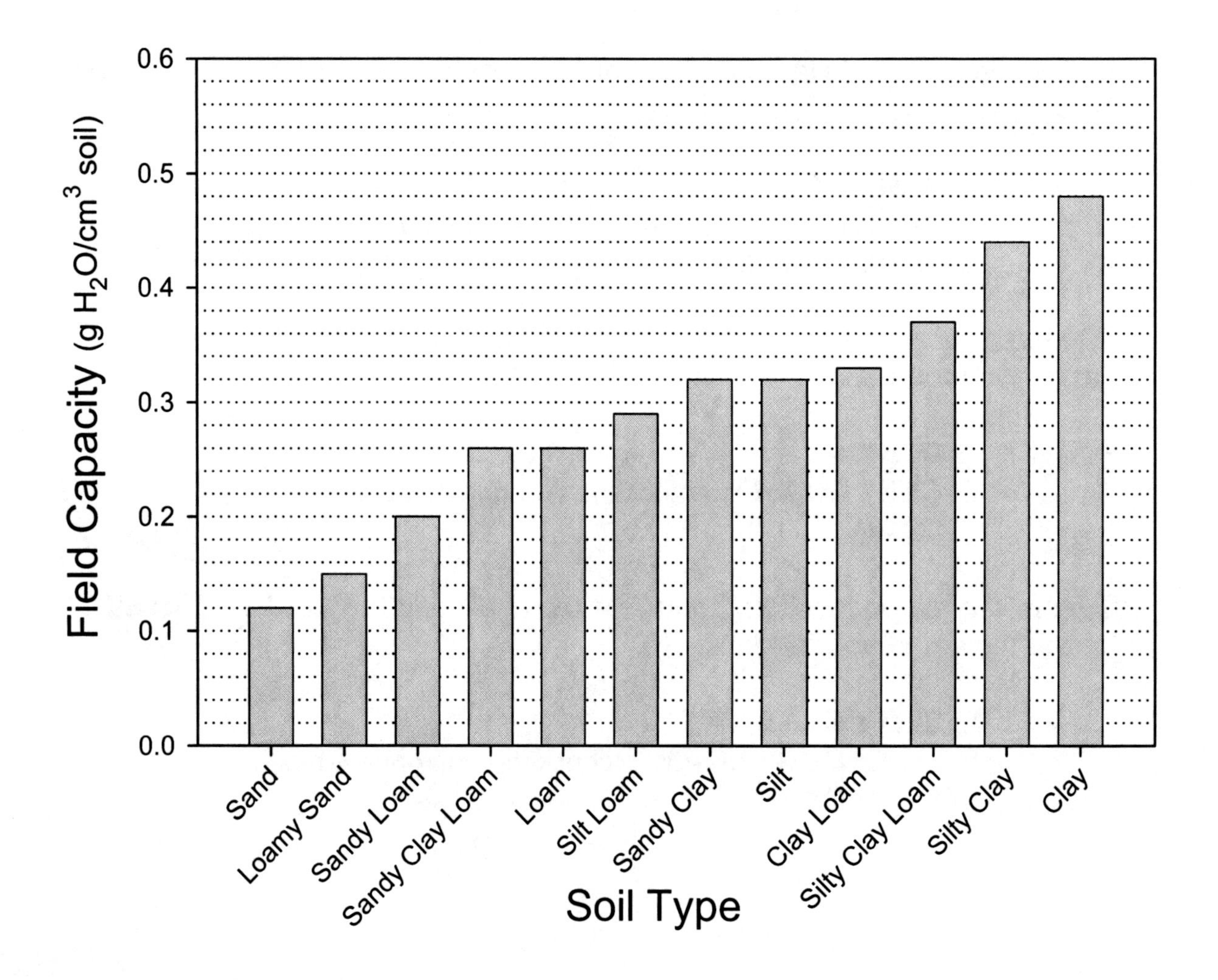

Procedure:

Determine from the above graph the Field Capacity of each of your soils.

Soil ____. Texture ______________________. Field Capacity ______ g H_20/cm^3

Soil ____. Texture ______________________. Field Capacity ______ g H_20/cm^3

Activity 3: Wilting point

Wilting Point refers to the amount of water in a soil below which plants, typically crop plants, will permanently wilt. As an example of the effect of soils on plants, consider the following graph that plots the *wilting point* of soils by the amount of clay present. Notice that soils high in clay have high wilting points.

Procedure:

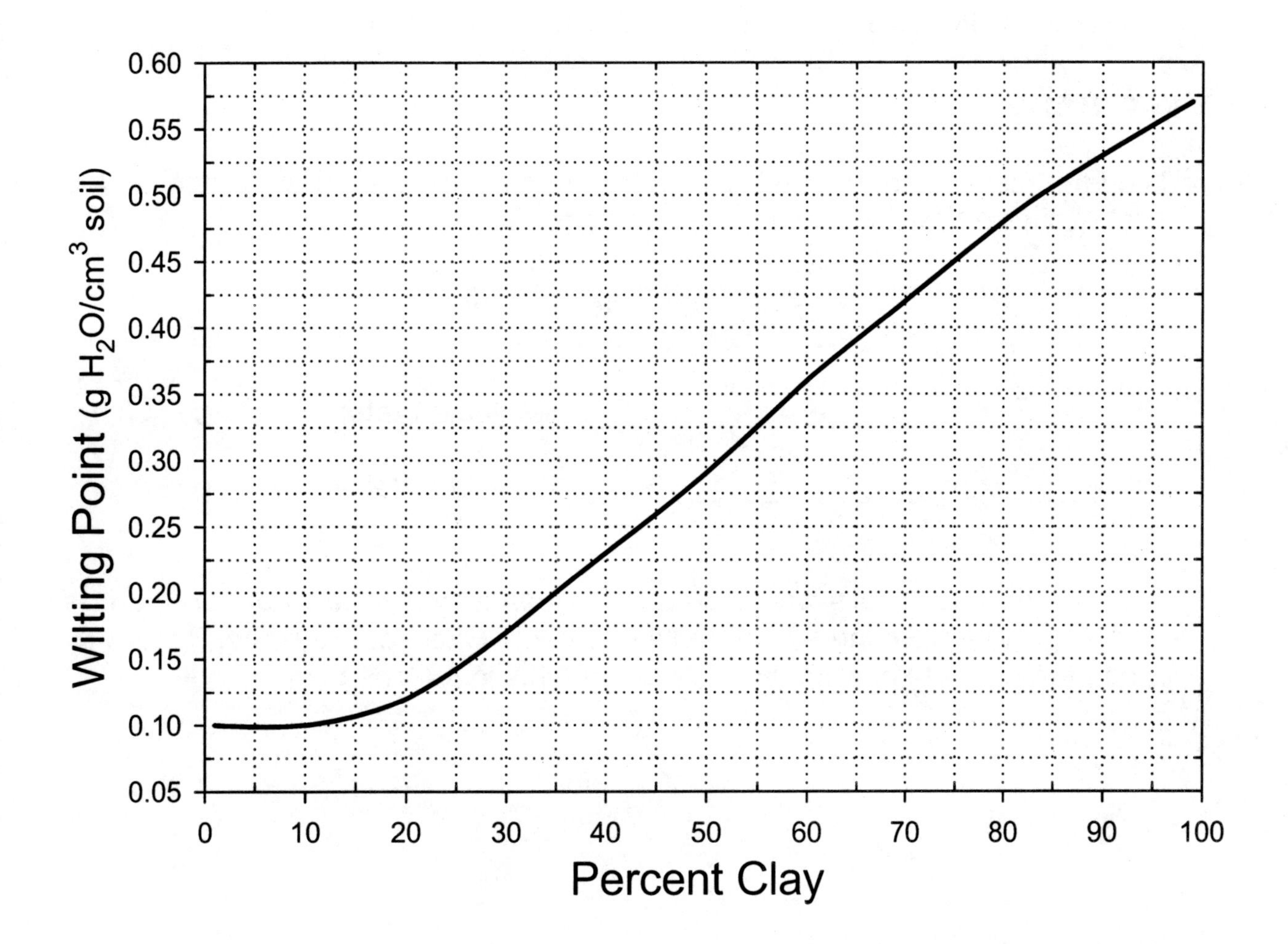

Use the Soil Pyramid to estimate an average percent of clay for each of the soils that you examined for texture. (Find the mid-point of your soil type on the Soil Pyramid and read off its percent clay. For example, for a Silt Loam soil, assume that the percent clay is about 15%). With the percent of clay you got from the soil triangle, determine the Wilting Point of each of your soils using the graph above.

Soil ____. Texture ____________________. Wilting Point ______ g H_20/cm^3

Soil ____. Texture ____________________. Wilting Point ______ g H_20/cm^3

Activity 4: Available water

So how much water is really available for plants to use? Some of the water present in a soil is not available for plant use. Once the wilting point is reached, plants begin to die because they can not extract any additional water from the soil. You can determine the amount of *available water* by subtracting the wilting point from field capacity.

Procedure:

Using the Field Capacity and Wilting Point of each of your soils, determine your soils' Available Water some time after a good rain:

Available Water = Field Capacity – Wilting Point

Soil ____. Texture ____________________. Available Water ______ g H_20/cm^3

Soil ____. Texture ____________________. Available Water ______ g H_20/cm^3

In which of your soils would corn or soybeans, which are not very good at extracting water from the soil, probably do best? Give your reasons. (Ignore the fact that we do not know the amount of nutrients in the soil.)

Other Physical Properties of Soil

Organic matter

Physically, soil is more than just texture. We have not examined, for example, the amount of *organic matter* in the soil. Organic matter is usually not considered to be part of soil texture; but along with silt and clay, it is important for holding onto water and storing some nutrients. You may have noticed that "potting soil" is very dark in color and is spongy. These properties are due to organic material. Other important components of soil include air, water, pH, and nutrients. All of these, however, are influenced by soil texture and the amount of organic matter present.

Time

Young soils are not the same as old soils. Consider the soils near the equator that have never experienced glaciation, they've been developing in place for millions of years. Soils in northern Iowa, however, were recently covered by massive glaciers and are therefore new soils. The older a soil is, the more time there has been for (1) erosion, (2) chemical reactions, (3) loss of nutrients dissolved in water moving out of the soil, and (4) biological activity.

Parent material

All soils develop from a *parent material*, most often the rock underlying the soil. If the parent material has many nutrients, usually so will the soil. If the parent material is acidic (pH less than 7), the soil will also be acidic. For example, soils derived from granite are acidic and low in nutrients and soils derived from dark-colored basalt are less acidic and have more nutrients. Soils derived from limestone are basic (pH greater than 7). Most soils have developed in place over the underlying parent material of bedrock. But many soils form from windblown material (*loess*) from elsewhere, like those of the Loess Hills of Iowa. Other soils form from parent materials moved into place by water (*alluvium*), glaciers (*glacial till*), and gravity (*colluvium*, like at the bottom of a mountain slope).

Topography

Topography refers to the shape of the land. The soils at the bottom of a slope are different than those at the top of the slope. For example, the soils at the bottom of a slope are deeper than those at the top because material is constantly being added due to gravity and water movement. Soils at the tops of hills and mountains are typically shallow. Further, lower slope position receive more water, due to run-off, and therefore have more biological activity.

Structure

You've probably noticed that soils often form "clods" when broken up. These clods, called *peds*, are also an important component of soils and make up a soil's *structure*. They may be large or small or may be absent altogether. Clayey soils are said to have *massive* structure, which means they have no discernible peds. Soil structure allows water to better percolate into the soils and forms channels through which plant roots can grow. Soil structure also helps to reduce soil erosion by binding soil particles together.

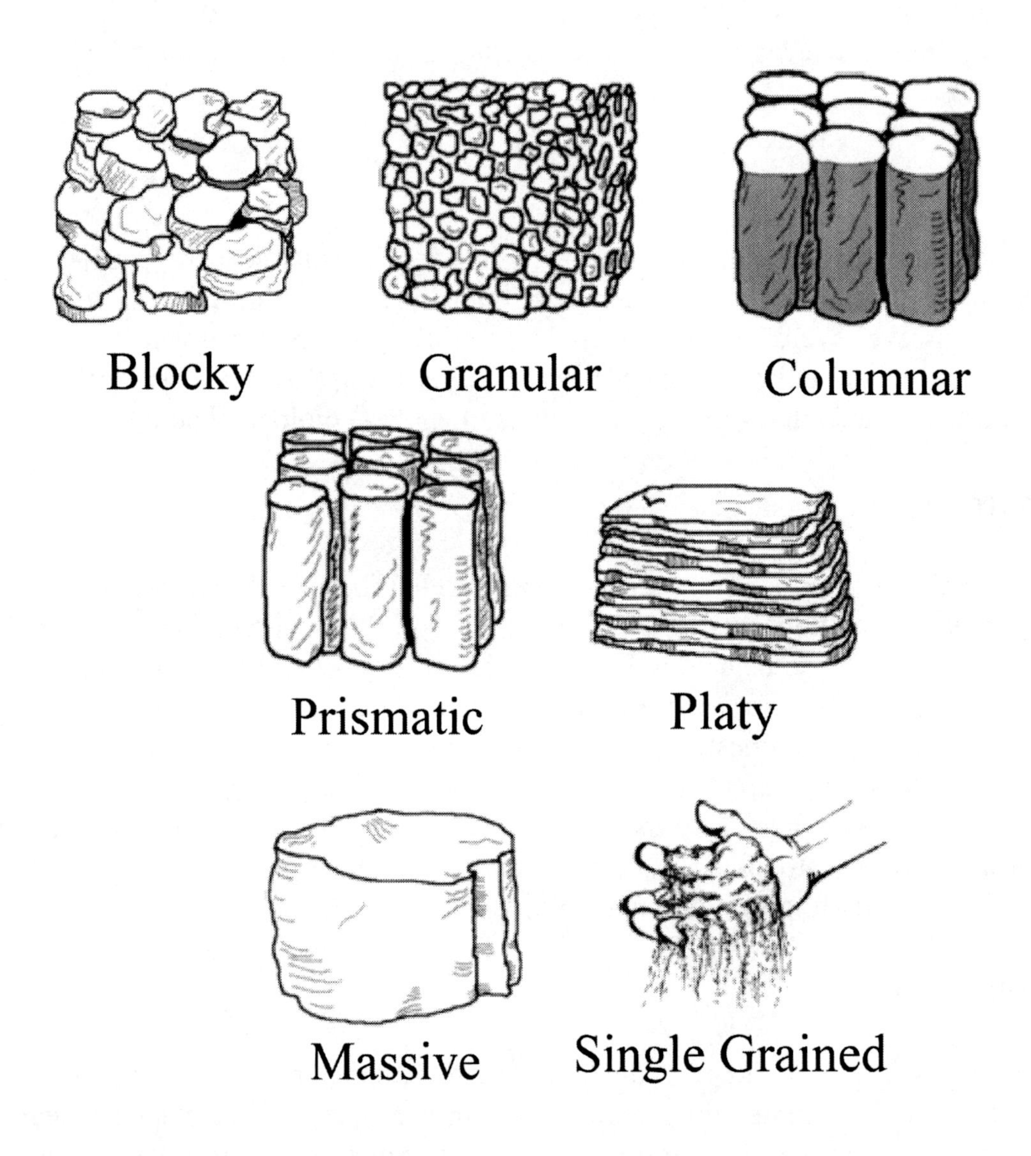

Diagrams used with permission of the NASA's Soil Science Education Web Page, soils.gsfc.nasa.gov

Climate

Climate determines, in large part, the kinds of organisms that can live in a particular area. Colder climates have fewer organisms and less biological activity. Warmer temperatures cause chemical reactions to proceed faster than they would in cold temperatures. Decomposition of biomass occurs at a faster rate in warm climates than in cold ones. More precipitation means more biological activity and more yearly production of plant and animal biomass. On the other hand, and especially in warm climates, increased moisture leads to increased decomposition of biomass by soil microorganisms.

Soils are "living"

In a way, soil is like a living organism: it grows and develops over time, has a history, is affected by its surroundings, has interrelated parts, and can be damaged or "killed." Furthermore, living organisms are an important component of all soils and have a powerful role in their formation. For example, worms form "casts" that contribute to a soil's nutrient content and structure – the mucous of their digestive tracts forms the soil into "crumbs". Moles, gophers, termites, and ground squirrels dig tunnels that allow air into the soil. Digging these tunnels also brings deeper soils to the surface.

Questions to consider

Sandy soils are often called *light* soils, while clayey soils are considered *heavy*? What do you suppose these terms are referring to? Hint: How would you like to dig a post hole in a soil that is pure clay?

Why would soils with a component of clay tend to have more nutrients?

Which kinds of soils would you expect to be most susceptible to wind and water erosion? Why?

How might pesticides damage a soil?

How might *acid rain* damage a soil? Hint: Acid rain is acidic because it contains many hydrogen ions (H^+). Most nutrients are also positively charged ions.

Why is it that clayey soils with worms are more fertile than those without them?

You notice that your crops are wilting yet the soil seems to still have water in it. What is the probable soil texture of your field? Explain your rationale.

In terms of changes to *soil structure*, why would plowing a field lead to increased soil erosion?

Part II: Biological Properties – The Living Soil

Have you ever thought about what is going on biologically beneath your feet? Probably not. Chances are, though, that a single heaping teaspoonful (about a gram) of healthy soil from where you live contains as many individual bacteria as there are humans on earth! Furthermore, this single gram of soil may contain 5,000 *species* of bacteria. Clearly the soil is much more than just a mixture of different chemicals and particles of varying size. It is almost literally alive.

Every gram of soil contains a diverse mixture of *microorganisms* like *bacteria*, *fungi*, and *actinomycetes* (a specialized form of bacteria). Without the activities of these microbes, plants would lack essential nutrients that would otherwise be tied up in dead organic matter. Microbes recycle nutrients from animal wastes and dead plants and animals In other words, their activities are important to *soil fertility* (the nutrient levels in the soil). Even the structure of the soil is partly determined by the organisms that live in it, as they and their products help to glue tiny particles of silt, clay, and sand into "crumbs" of soil.

The activities of soil microorganisms are an important part of the *carbon cycle*. Microbes use the organic compounds in decaying organic matter as an energy source. They burn these compounds to supply themselves with energy in the process of *respiration*. Carbon dioxide (CO_2) released from *respiration* enters the atmosphere. Plants and other photosynthetic organisms then use this carbon dioxide to make new carbon-containing compounds in the process of *photosynthesis*.

Soil microbes are also the major contributors to the *nitrogen cycle*. Some microbes break down proteins to release ammonia, which can be used by plants as a source of nitrogen. Other microbes can convert, or *fix*, atmospheric nitrogen (N_2) into forms that are useable by plants. Plants and animals are incapable of nitrogen fixation on their own.

In today's lab, you will observe prepared laboratory cultures of microorganisms from a single gram of local soil. You will estimate the diversity of different types of microorganisms on your plates and also estimate the numbers of microorganisms in a gram of soil.

Definitions

actinomycetes: Soil bacteria that form branching strands of cells and can form spores. Actinomycetes are the source of many antibiotics, including streptomycin and tetracycline. The actinomycetes can degrade resistant organic material, such as rubber products, and are also responsible for the smell of moist earth.

microorganism: An organism that is not visible without a microscope.

photosynthesis: The process where green plants, some non-green algae, and cyanobacteria (blue-green "algae") fix atmospheric CO_2 into organic molecules using light as an energy source.

respiration: Using organic compounds to provide energy. Respiration releases CO_2. *All* organisms respire, including all plants.

The number of microorganisms in even a gram of soil is usually very (very!) large. Spreading an undiluted soil solution on nutrient medium in a petri plate results in a huge number of overlapping, indistinguishable *colonies* (groups of one species of microbes). Diluting the sample in known amounts of sterile water before spreading it on the plates solves this problem. The figure below illustrates the dilution procedure used to make the microbe plates for Part II's procedures.

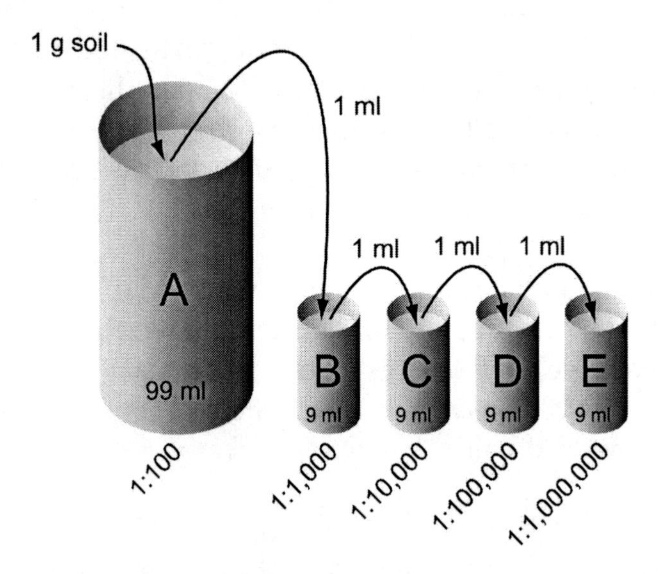

Series of dilutions in sterile water and dilution factors for sampling soil microorganisms.

Activity 1: Microbe Diversity

Procedure:

1. ☐ In groups of 2-4 people, obtain 2-3 nutrient plates from your instructor. These plates were made by diluting 1g of soil with sterilized water to make a solution of 1:1,000,000,000 (see figure above), and then inoculating the plate with 0.1 ml of this diluted sample. The plates were then left in an incubator at around 30 degrees Celsius for 1-2 weeks.

Precautions: Do NOT open the plates – keep them closed to prevent contamination. Also, treat all microorganism as if they are pathogens. The risk of encountering anything dangerous in you soil sample is, however, very low.

2. ☐ Invert your covered plates on the lab bench surface or a sheet of white paper, which ever works best for you. You will be looking through the medium to see the microorganisms.

3. ☐ Choose one plate and count the number of different bacteria and the number of different fungi that you see. Record these on the data table below.

Different species will be different sizes, shapes, textures, heights, margins, and/or colors. Bacterial colonies usually look slimy, shiny, chalky, or like a drop of wax but are not fuzzy. Fuzzy stuff is fungi (technically the fungal hyphae or filaments).

Some microbes secrete chemicals that inhibit the growth of nearby bacteria. Carefully look at the plates to see if there are any *zones of inhibition* -- area/space around colonies where other bacteria are not found. Some of these chemicals are used in pharmaceuticals to make antibiotics (drugs that kill bacteria infections).

You may wish to list descriptions of the types of microbes you find on a separate piece of paper in order to keep track. Your descriptions should be fairly specific, e.g., chalky-white with darker center, highly peaked colonies, etc.

Microbe Type	**Plate 1**	**Plate 2**	**Total**
# of Different **Bacteria**			
# of Different **Fungi**			

4. ☐ Now repeat Step 3, but this time looking at the second plate. Then add up your results to find out the total number of different bacteria and the total number of different fungi in the soil sample.

Activity 2: Estimating Microbial Population Numbers

As you may have noticed from the previous activity, a typical gram of soil contains much more bacteria than actinomycetes (a fungi-like bacteria, some of which are used to make antibiotics) or fungi. Fungi are the least abundant.

In Activity 2 we will count the number of bacteria and fungi on nutrient plates inoculated with 0.1mL (that is 1/10 of a mL -- or 100µL) of a 1g soil sample that had been diluted to 1:100,000. We will then use a formula to estimate the total number of bacteria and fungi that can be found in 1 gram of soil.

Procedure:

1. ☐ From your instructor, get *one* bacterial plate and *one* fungi plate that contains between 30 and 300 colonies of organisms.

2. ☐ Invert the covered plate on the lab bench surface or on a sheet of white paper, whichever works best for you. You will be looking through the medium to see the colonies.

3. ☐ Use an erasable marking pen to "tick off" the bacteria colonies as you count them. Count *all* the colonies of bacteria on your plate. Record the number of colonies in the table below.

4. ☐ Using a different colored erasable marking pen, tick off the fungi colonies as you count them. Be sure to count every fungi colony. Record the number of all fungi in the table below.

5. ☐ Now calculate the number of microorganisms (of bacteria or fungi) per gram of soil using this formula:

Microorganisms per gram of soil = Colony Count × Dilution Factor × 10

For example, if you found 100 bacterial colonies on the plate with the 1:100,000 dilution, the number of bacteria per gram soil is 100 × 100,000 × 10 = 100,000,000 or 10^8. You have to multiply by 10 because only one-tenth of one milliliter of soil solution was spread on the nutrient plate to make it.

Table 1: Estimating Microbial Population Numbers

Microorganism Type	Dilution Factor	Colony Count	Number of microbe per g soil (count × dilution factor × 10)
Bacteria	1:100,000		
Fungi	1:100,000		

6. ☐ CLEAN UP: Using a damp paper towel, wipe off the colored ticks you made on the plate and then return all plates used in Part II to the instructor.

Questions to Consider

How many different kinds of bacteria and fungi were in your sample?

How could you tell the difference between different kinds?

Were the different kinds of microorganisms on your plate represented equally, or were one or two kinds dominant? Why?

Did your class observe greater diversity in bacteria or fungi?

Culturing techniques, like the ones you performed in this lab are known to underestimate the actual numbers and diversity of microorganisms in the soil. Suggest a possible reason for this.

Actinomycetes produce some of the most important antibiotics used by people. Why do you think actinomycetes naturally produce antibiotics?

How could you tell if one of the colonies on your plate produced an antibiotic? (Drawing a picture is an effective way to answer this question).

Part III: Soil Animals

You have seen how many microorganisms live in a small sample of soil. As you can imagine, they do not live there alone. *Soil microorganisms exist within a web of interacting organisms, of which they just happen to be the smallest.*

Many, many different kinds of organisms make their living in the soil. Some of them, such as tiny roundworms and amoebas, use microorganisms as food. Other soil *invertebrates* (spineless animals), earthworms for example, assist the microbes in the business of recycling organic compounds by breaking down plant and animal waste and other dead material into smaller parts. Some invertebrates feed on or parasitize the living roots of plants. (These are some of the best-studied soil animals because their activities cause serious economic losses to farmers.) There are also predatory soil invertebrates that prey on the other kinds!

From our perspective, the soil looks mostly solid and impenetrable. But if you could shrink down to the size of a soil invertebrate you would see a labyrinthine system of tunnels and spaces of varying sizes. The smallest pores in the soil and the spaces inside soil crumbs are occupied by microorganisms. The larger pores are home to roundworms and tiny insects. Earthworm tunnels and the spaces made by plant roots, besides letting air and water move into and through the soil, are like highways for soil animals.

Light doesn't penetrate very far into the world of soil animals. Consequently, animals that spend their entire lives in the spaces between soil crumbs often have very reduced eyes, if any. They rely on sophisticated senses of touch and the ability to sense variations in the concentrations of chemicals (a kind of sense of "smell") to avoid predators and obtain food.

Finding a mate presents special difficulties for tiny soil animals. The males of some species of mites (relatives of spiders) and springtails (insect-like, wingless soil organisms) have given up on *finding* females. Instead they leave packets of sperm attached to delicate stalks in soil tunnels. Females that happen upon the right kind of sperm packet use the sperm to fertilize their own eggs.

In this portion of the lab, you will closely examine a sample of topsoil and identify and record the kinds and numbers of soil invertebrates that you find. You will compare different techniques for sampling soil animals, and you will compare your findings with those of other student groups who looked at soil obtained from different locations.

Activity: Soil Invertebrates

In this activity you will use several procedures to extract different types of soil invertebrates out of soil samples. At the end of these procedures are data forms to help you summarize what you found and the methods you used to find them. Afterwards you will visit other groups in the lab to see what they found and create a new table that summarizes soil invertebrate diversity from several class samples.

Procedure 1: Direct Observations

1. ☐ Work in groups of 3–4 students. Make sure you all have a chance to see and understand each result.

2. ☐ Look over the "Illustrated Guide for Identifying Soil Animals" at the end of this Lab. Familiarize yourself with some of the most common kinds of soil animals before you start to look for them.

3. ☐ Gather the following supplies:
   - ☐ sheets of newspaper
   - ☐ a sheet of white paper
   - ☐ a sheet of black paper
   - ☐ a kitchen sieve
   - ☐ 2 hand lenses or magnifying glasses
   - ☐ paint brushes
   - ☐ petri dishes

4. ☐ Get a sample of about 500 ml of soil from one of the boxes or buckets. Record the soil source (e.g., garden, woodland, prairie, etc.) on the Data Form on pages 20-21.

5. ☐ Pour the soil sample onto the newspaper. Carefully observe using a hand lens. Do you see any "panic-stricken" tiny animals trying to bury themselves in the soil? Use a paint brush to catch specimens.

6. ☐ Place specimens in a petri dish and observe with a dissecting microscope. You may need to kill some of the more active animals with ethanol from the wash bottle in order to observe them closely.

7. ☐ Use the "Illustrated Guide for Identifying Soil Animals" on pages 25–29 to find out what kinds of organisms you have found.

8. ☐ Record all your finds in the Data Form (pages 20-21), checking the box for Direct Observation (DO) in the Technique column.

9. ☐ Now divide the soil sample into two piles. Sift one of the piles onto white paper. Use a hand lens and paint brush to catch specimens, identify them using the dissecting scope, and record your data. (The technique is "WP" for White Paper.)

10. ☐ Sift the other pile onto black paper. Catch, identify, and record. (Check "BP" for technique.)

11. ☐ CLEAN UP: Place soil in the "used soil" box.

Procedure 2: Tullgren-Berlese Funnels

This technique takes advantage of the tendency of soil invertebrates to move away from light and heat. The soil sample is placed, intact if possible, in a screen-lined funnel. A light-bulb provides light and heat to drive soil animals down the funnel into a container of alcohol below.

1. ☐ Obtain a small jar of alcohol from one of the Tullgren-Burlese funnels. Use a jar from the same "soil source" as before.

2. ☐ Swirl the jar gently to suspend any animals that have sunk to the bottom. Pour the alcohol into a petri dish and observe using the dissecting microscope. Please handle the specimens gently so that they can be observed by students in other lab sections.

3. ☐ Identify and record your finds in the Data Form, noting that the technique is Tullgren-Burlese (TB) funnel.

4. ☐ CLEAN UP: Return the alcohol and preserved animals to the jar.

Procedure 3: Baermann Funnels

Most soil nematodes (small roundworms) are too small to be observed with a hand lens or dissecting microscope (see page 25 for pictures and more informaion). Tullgren-Berlese funnels are ineffective for capturing them because nematodes respond to light and drought by entering a "resting state" that resists drying out, not by fleeing. To observe nematodes, we will take advantage of their instinct to move from dry soil to wet soil. A Baermann funnel (or nematode trap), suspends a small sample of soil held by fine-mesh screening above water so that the soil is just touching the water. Nematodes move from the dryer upper soil through the wetter soil and into the water. They collect at the point of the funnel. Soil amoebas and other protozoans can also be found in the water in a nematode trap.

1. ☐ Use a plastic pipette to carefully remove a drop of the water (and soil particles) from the point of a nematode trap. Again, stay with the same soil source as before.

2. ☐ Place the drop on a clean microscope slide.

3. ☐ Observe the slide using the compound microscope. Start with the lowest power lens and move to higher power as you find living organisms.

4. ☐ Look for tiny, transparent worms that move with a whip-like motion. These are nematodes. They are usually near soil particles.

5. ☐ Record on the Data Form under "Tally" the number of nematodes you found in your wet mount. Nematodes are best examined at 100× total magnification. Check NT (Nematode Trap) for Technique. (See below for an explanation of the Data Form.)

6. ☐ Increase the magnification and adjust the light (iris diaphragm) to view the "insides" of the worms. At higher magnification (400×), you may also see smaller, fast-moving "fuzzy blobs" that zig-zag across your field of view. These are single-celled *protozoans* that feed on soil bacteria. Record these, too, on the Data Form under NT.

7. ☐ CLEAN UP: Put the used slides in the used slides container. Carefully return the nematode funnel to the side counter. Wipe off microscope lenses with lens paper.

Data Form Instructions (form on next page)

Description of Organism: Include here the general type of animal (e.g., mite, springtail, centipede) and any descriptive terms (color, surface texture) that help you to recognize it. Example: fuzzy, red mite.

Size: Record "Size" in three classes: Large (>1 mm), Small (<1 mm), or Micro (microscopic)

Technique: Check the box for each observational technique with which your group observed this specific kind of organism.

DO = Direct Observation
BP = Black Paper
WP = White Paper
TB = Tullgren-Burlese Funnel
BF = Baermann Funnel

Data Form 1 Soil Source: ____________________

			Technique				
Description of Organism	Size	Tally	D O	B P	W P	T B	B F

Data Form 2 Soil Source: ____________________

Description of Organism	Size	Tally	Technique				
			DO	BP	WP	TB	BF

Procedure 4: Group Comparisons

1. ☐ Use your Data Forms to fill out Table 1 below.

Table 1: Group Soil Invertebrate Summary

Category	Results of Your Group's Sampling	
	Soil Source:	Soil Source:
Total Number of Soil Animals		
Number of Different Kinds of Soil Animals		
Most Abundant Kind of Soil Animal		

2. ☐ Visit other groups and take a look at the organisms they found in their petri dishes. Then ask someone in that group for their Table 1 data. Write this information in Table 2. Make sure you enter your data in a row corresponding to the soil source you sampled. To calculate an average, add up all the values in one column and then divide by the number of groups you have.

Table 2: Class Soil Invertebrate Summary

Soil Source	Group	Total Number of Soil Animals	Number of Different Kinds	Most Abundant Soil Animal
	1			
	2			
	3			
	4			
	Average:			
	1			
	2			
	3			
	4			
	Average:			

3. ☐ **CLEAN UP:** Place the live soil critters from your petri dishes in the critter jars at the front of the room. Dump out any dead ones in the garbage. Then clean, wash, and dry all petri dishes, brushes, sieves, and tweezers. Brush off the black and white paper and place these back on your table. Toss out any used newspaper and used plastic pipettors. Lastly sweep all soil off your table and neatly arrange lab supplies on your table.

Questions to Consider

How do springtails escape their predators?

What is the fundamental food source for soil animals?

Was there a difference in the kinds of organisms you could find, depending on the color of the background paper?

Based on the kind of device used to collect nematodes, what environmental conditions do nematodes prefer?

What do most soil invertebrates do when the weather is hot and dry?

Since soil invertebrates live in the pores and tunnels in the soil, what happens to them when the soil is tilled? Would you expect to find that the populations of soil invertebrates would be higher, lower, or unchanged in response to tilling?

Did the total numbers and diversity of organisms differ between soil sources? (Refer to your class data.) Can you link any differences to characteristics of the soil sources, e.g., type and amount of organic matter or level of disturbance (such as digging).

Many nematodes consume soil bacteria. Could you use the numbers of nematodes as an index of microbial population size?

Would you expect to find more soil invertebrates in a field grown with synthetic fertilizers and pesticides or an “organic” field, where compost and manure are used to maintain fertility and natural pest-control is practiced? Explain your answer.

Illustrated Guide to Common Soil Animals

Illustrations by Chris Schulte

Worms

Earthworms have segmented bodies without appendages. They move by contracting and extending their segments and ingest soil microbes and decaying plant material. As they eat and defecate their way through the soil, worms help to create soil crumbs and improve water penetration and soil aeration.

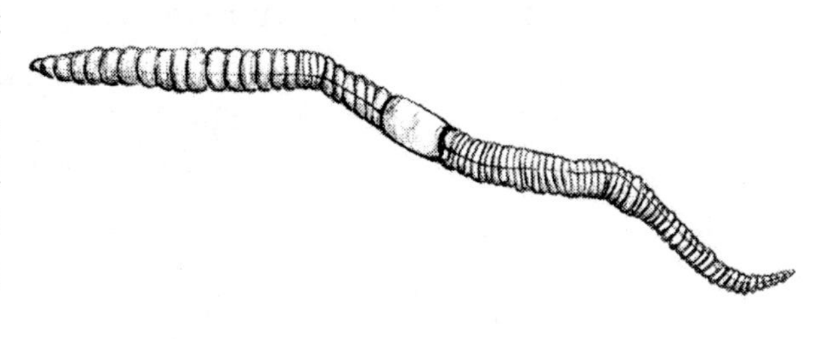

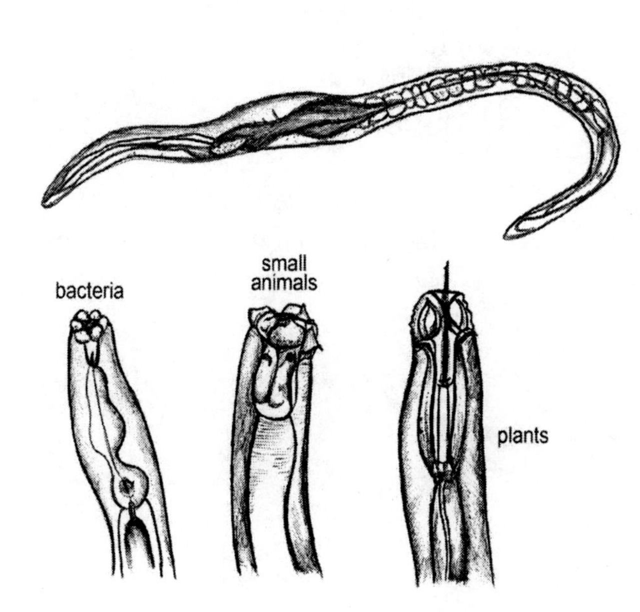

Nematodes, which are small roundworms, lack segments and move by whipping from side to side. They are some of the most abundant soil animals. If planet earth were to suddenly become invisible, except for its nematodes, the outline of the planet would still be visible from space due to the abundance of nematodes in the soil! Most nematodes are too small to be seen without magnification.

Different kinds are specialized to feed on different food sources.

Useful References:
Eisenbeis, G. and W. Wilchard. 1987. *Atlas on the Biology of Soil Arthropods*. Springer-Verlag, Berlin.
Nardi, J.B. 2007. *Life in the Soil*. University of Chicago Press.
Rhine, R. 1972. *Life in a Bucket of Soil*. Lothrop, Lee & Shepard, Co., New York.
Schaller, F. 1968. *Soil Animals*. Univ. Michigan Press, Ann Arbor.

Insects and Insect-like Animals
(animals with six legs)

Springtails, or Collembolans, are extremely abundant soil animals. Because they have six legs, they were long-thought to be insects. We now know that they are not insects, mainly based on the fact that they have internal, rather than external, mouth parts. Most are very small, less than 5 mm long. They feed on dead plant matter and fungi and are themselves and important source of food for ants and other predators. Surface-dwelling species have a long "spring" that helps them to escape predation. Springtails deeper in the soil have shorter springs; at the deepest levels, the spring is absent. Eyes also decrease in size with depth until they may be missing altogether.

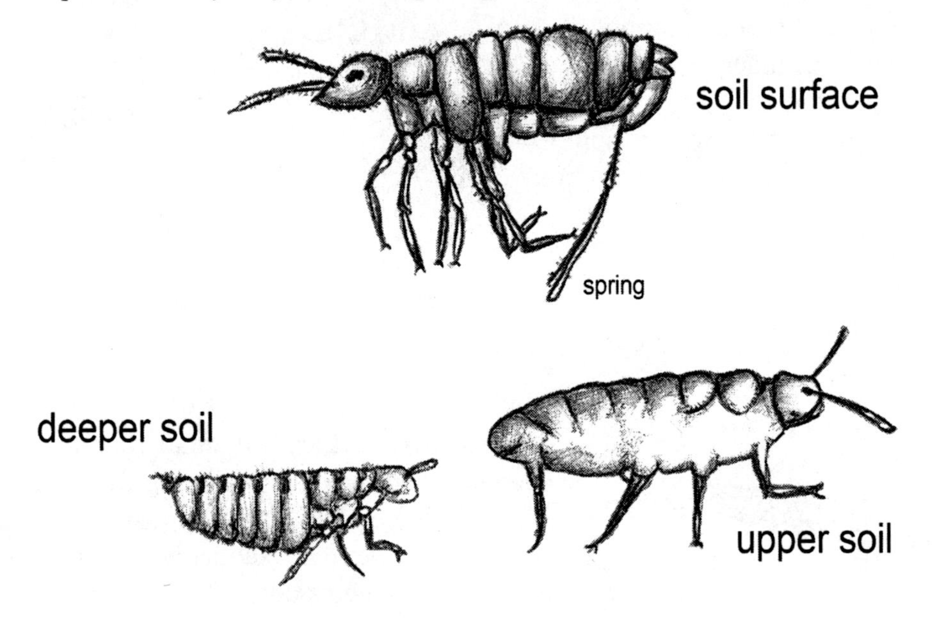

Doubletails, like the springtails were once considered insects but have internal mouth parts. Some feed on decaying wood or leaf litter; others are predatory.

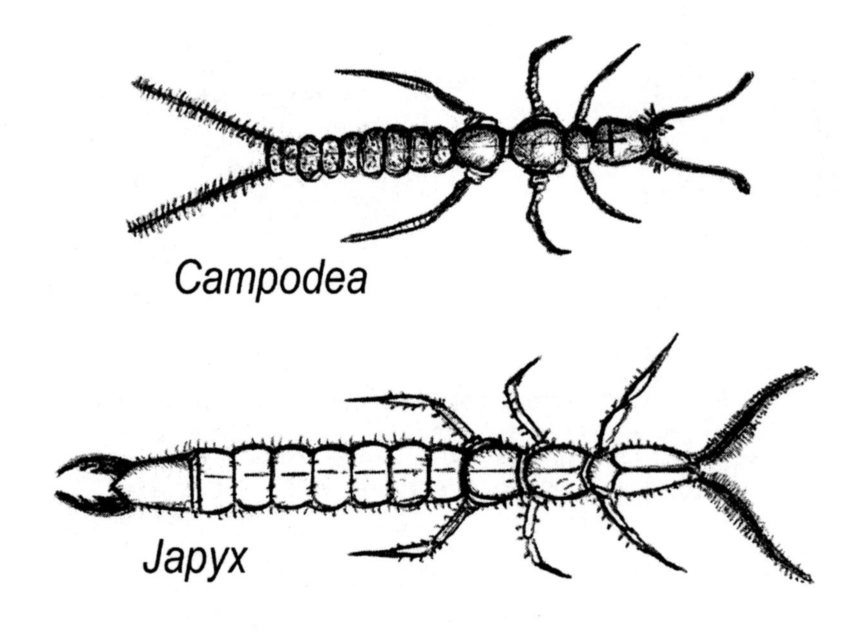

Mature **beetles** have two pairs of wings, of which the outer pair is hardened and protective. The larvae (grubs) lack wings. The larvae may resemble the primitive "insects" like springtails and doubletails but are usually much larger. Many kinds of beetle larvae live and develop in the soil. Some eat living plant roots, others feed on dead organic matter, and some are predatory.

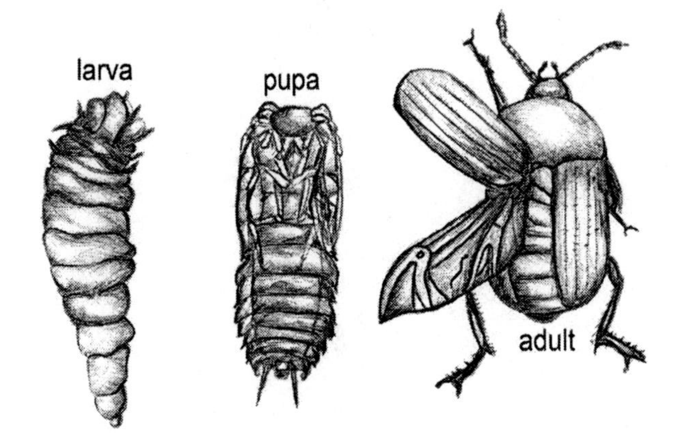

June beetle. The larvae feed on plant roots.

Earwigs are important soil insects. They feed on various substrates, like live and dead plant parts, fungi, insects, and spiders. Earwigs have pincers (for defense) and reduced wings. Female earwigs guard their eggs and young.

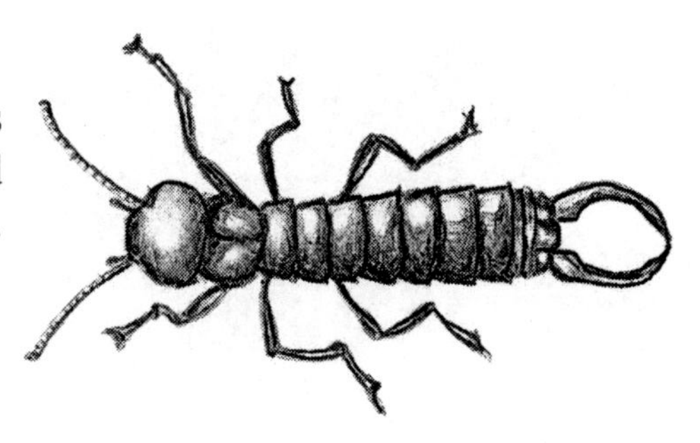

Soil-dwelling **cockroaches** feed on rotting plant material.

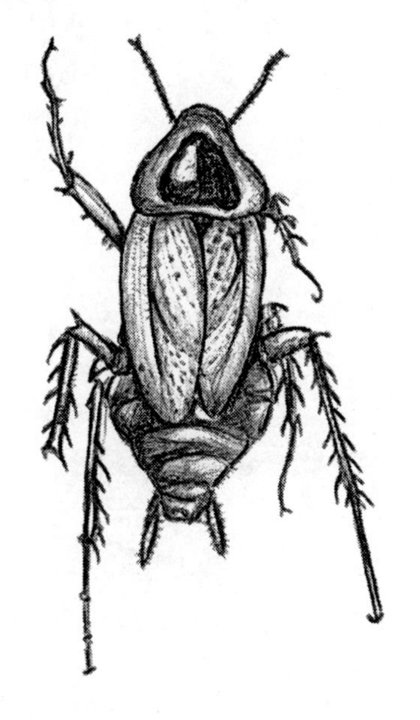

Field **crickets** drag fresh plant material into their burrows to feed.

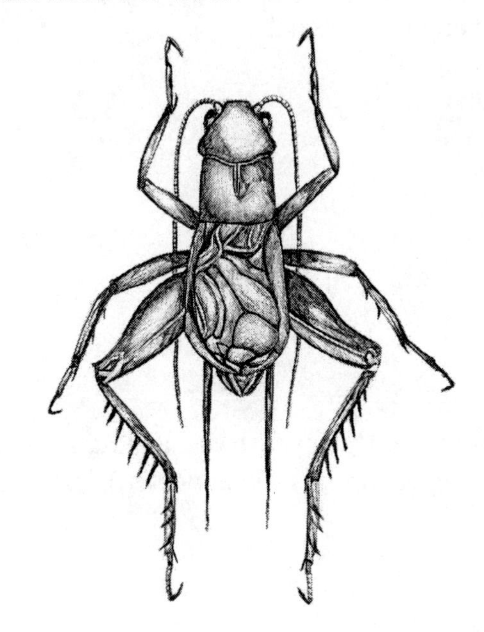

Animals with 8 or more legs

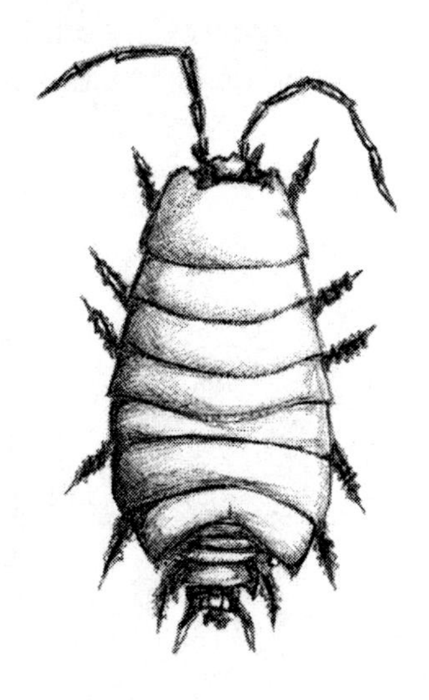

Isopods (wood lice, pill bugs, roly-polies) have 14 legs and feed on dead plant material. They can roll up inside their armor for protection from predators.

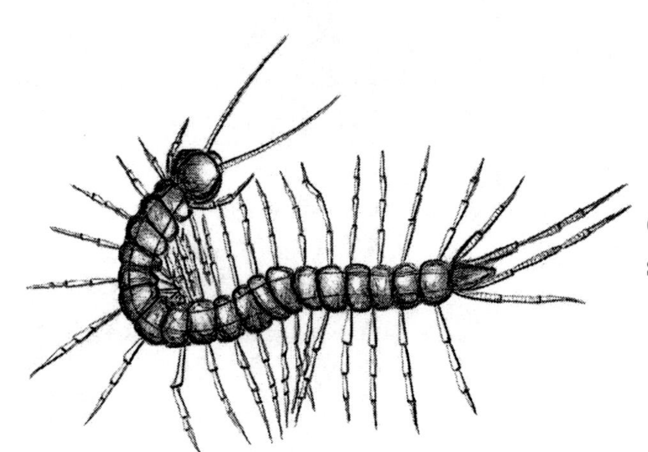

Centipedes have one pair of legs per body segment and are predators.

Millipedes have two pairs of legs per body segment and usually feed on dead plant material.

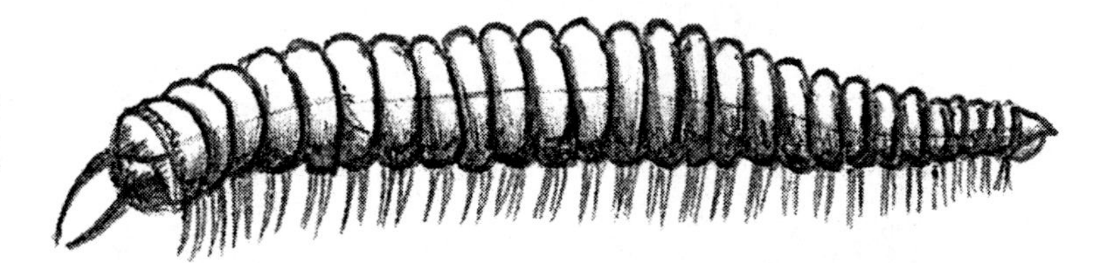

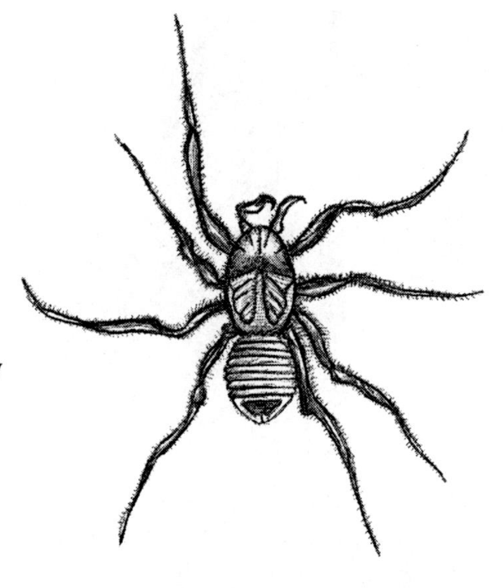

Spiders, and their relatives, have eight legs and two main body parts (cephalothorax and abdomen). They are predators.

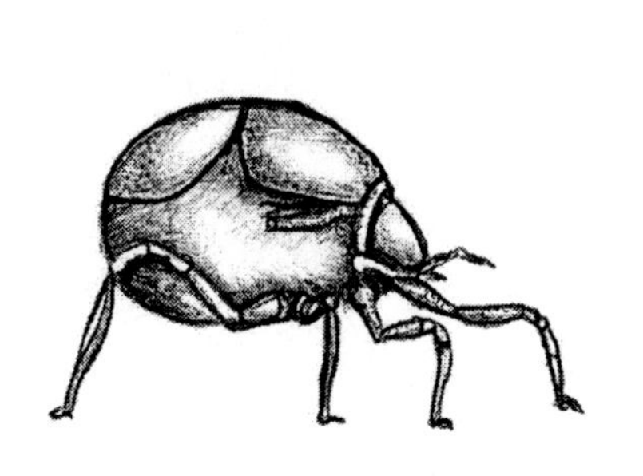

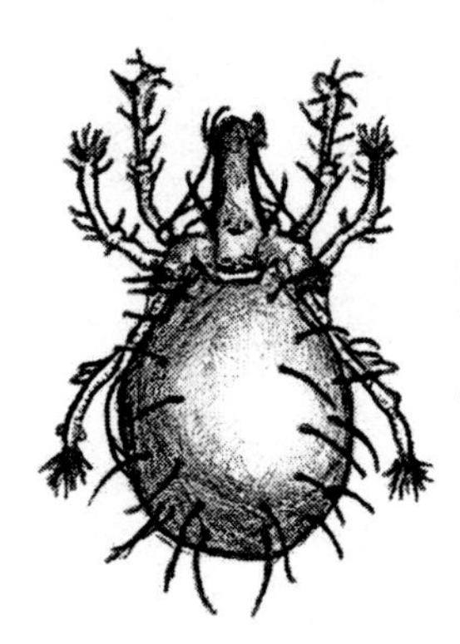

Soil **mites,** like their relatives, the spiders, scorpions, and pseudoscorpions, have eight legs and two main body parts. Most feed on dead organic matter.

Pseudoscorpions, the false scorpions, like their poisonous relatives, have two pinchers to grab prey organisms. Poison glands are found in the movable half of the pincers. The poison is used to immobilize their prey. (Don't worry, they won't hurt you.) These 8-legged animals are predatory on other soil animals, especially small insects and mites. In order to eat their prey, these animals "spit" a digestive liquid on their prey. Once the prey is liquified, it is ingested.

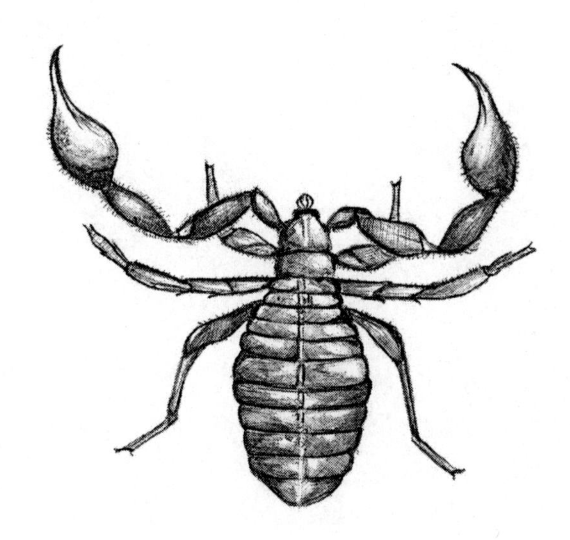

Name: ______________________________ Instructor's Name: _____________

Normal Lab Day and Time: _______________

Natural World Lab Response Sheet

Lab 9: Soil

1. What was the main idea behind today's lab?

2. Looking the 100ml graduated cylinder provided, determine the soil type of the sample using the *Soil Pyramid. Show your calculations.*

3. Which of the following statements about soil invertebrates is/are true?
(Use your lab experiences and readings from Part III to answer)
   A) In general, soil invertebrates move toward light and heat.
   B) Most soil invertebrates are longer than 10mm (1cm).
   C) Most soil invertebrates are brightly colored ad conspicuous.
   D) Some male soil invertebrates leave sperm packets in soil tunnels instead of trying to find females.

4. Explain why soil with high microbial activity would probably have a high degree of fertility (rich in nutrients)? (See Part II readings for help)

5. Rank today's lab from 0 (poor) to 10 (excellent). <u>WHY</u> did you choose this rating?

LAB 10: AQUATIC TOXICOLOGY

We're all connected, one way or another, to the same great global cycle of water as it moves from the atmosphere to the ground (and oceans), from the ground to the plants, from the plants to the animals, and eventually back to the atmosphere by the process of *evapotranspiration* (evaporation from the oceans and the ground plus transpiration from plants). In a typical ecosystem about 70% of the precipitation that falls on it returns directly to the atmosphere by evapotranspiration from plant leaves! In 1977, Congress passed the Clean Water Act that prohibits the discharge of pollutants at toxic levels into oceans, lakes, streams, and wetlands. Thousands of chemical compounds and elements are daily discharged into the environment from industrial, agricultural, and residential sources. In the past there was only limited or nonexistent information regarding the levels at which human-caused waste products become toxic when introduced into ecosystems. We now have a much better idea of the extent of the problem.

In order to effectively protect both the health of ecosystems and of the public, it is important for scientists to assess the potential hazards of environmental pollutants. Most pollutants discharged into the atmosphere or applied to the soil eventually end up in fresh-water or marine ecosystems. Consequently, aquatic organisms have proven to be cost-effective and sensitive indicators of the amount of toxic pollutants being deposited on the landscape. We will use *Daphnia magna*, a small freshwater crustacean – a shrimp and crab relative – as a bio-indicator of pollution levels. *Daphnia magna* (illustrated below along with another species, *Daphnia pulex*) feeds on microscopic algae and single-celled animals and is often a major food source for small fish.

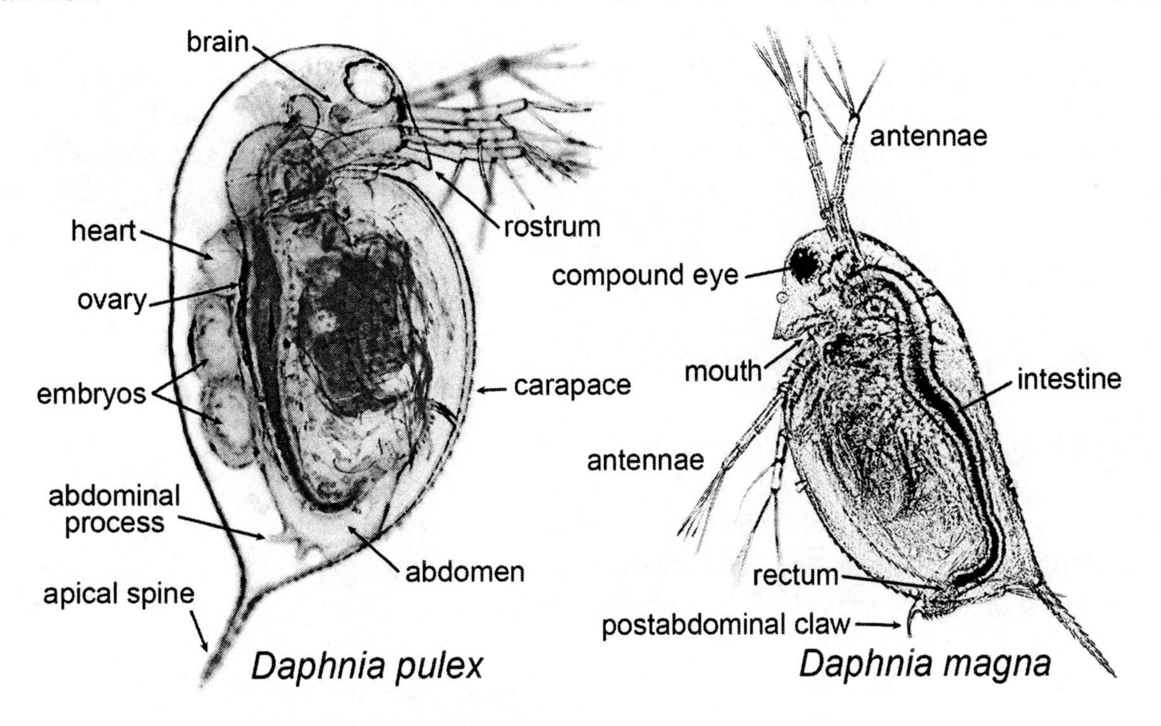

Toxicity tests are used to determine the concentration of a pollutant that is lethal to 50% of test organisms over a specified time period. This concentration is referred to as the Median Lethal Dose or LD_{50}. Small fish, mussels, and water fleas (*Daphnia*) are often used to give a general indication of the safety of water sources. Changes in the health, mortality, or behavior of these *indicator organisms* is a warning of the presence of toxic hazards – a kind of "canary in the coal mine."

In this lab, you will determine the LD_{50} for *Daphnia* exposed for 30–48 hours to varying concentrations of zinc (a metal related to cadmium and mercury). Sources of zinc in aquatic systems include mining wastes and untreated sewage. Although zinc can have serious consequences for aquatic ecosystems, it is not toxic to us at the concentrations we will be using.

Activity 1: Observations of *Daphnia*

Procedure:

1. ☐ Use a plastic transfer pipette to catch one or two large *Daphnia.* Place them in a drop of water on a depression slide. Look for individuals that have embryos beneath the "skin" of their backs. These look like dots.

2. ☐ Observe the *Daphnia* using the scanning (4× or 5×) objective lens of a compound microscope or use a dissecting microscope, whichever is available.

3. ☐ Watch the swimming motion of the *Daphnia*. In Procedure 2, you will use movement as an indication that an individual *Daphnia* is alive.

4. ☐ Are there embryonic *Daphnia* inside some of the adults? In Activity 2 it will be important to use the smaller, non-gravid (non "pregnant") *Daphnia*. That way you will avoid finding more, rather than fewer, *Daphnia* at the end of the experiment!

Activity 2: The LD_{50} Test

Procedure 1: Setting up the LD_{50} Test

1. ☐ Work in groups of 3 to 4 students.

2. ☐ Label 4 plastic cups with your group's name, lab time and day, and zinc concentration. You will use four concentrations of zinc, given as milligrams (mg) of zinc per liter (L): 0.1 mg/L, 0.5 mg/L, 1.5 mg/L, and 4.5 mg/L.

3. ☐ Use a graduated cylinder to add 50 ml of the appropriate solution to each labeled cup. IMPORTANT: rinse the graduated cylinder with distilled water after each use!

4. ☐ Use a plastic pipette to transfer 10 **small** (non-pregnant) *Daphnia* to each of the test cups. BE CAREFUL not to touch the zinc solution in the cups with the pipette. If you do, some of the zinc will contaminate the pipette and you will need to throw out that pipette. Also, place as little of the liquid from the *Daphnia* container in your cups to avoid changing the concentration of you solution. Throw out any used pipettes after you have completed this part of the lab.

5. ☐ Move the four test cups which now contain *Daphnia* to the area marked for your lab section. They will remain undisturbed for 30–48 hours. **You must come back to make observations 30 to 48 hours later (Procedure 2). Make a note in your calendar or planner to do this!**

Procedure 2: Observe results 30–48 hours later

1. ☐ Count the number of dead *Daphnia* in each of your test cups.

   Note: *Daphnia* are considered to be alive if they are moving, even slightly. Observe each individual for several minutes to see if it moves at all. Dead *Daphnia* often have a whitish "ghostly" appearance.

2. ☐ Record the numbers in Table 1, which is located below.

3. ☐ Repeat the observations on a set of cups from another group in your class. Again, record these numbers in the table.

4. ☐ Calculate the average number dead *Daphnia* for each test concentration: To find the average number of dead *Daphnia*, add the numbers of dead in both cups at each test concentration and divide by two. Enter the results in Table 1 as the Average Number Dead.

5. ☐ Calculate the average percent mortality for each test concentration: divide the average number dead by ten (the original number of *Daphnia* in each cup) and multiply by 100. Record the results in Table 1.

6. ☐ Plot the average percent mortality for each concentration of zinc on the graph below. Draw a ***straight line*** (NOT a zig-zag "dot-to-dot" line) that best fits the data through the scattered points on the graph.

7. ☐ Find 50% Mortality on the *y* axis. Draw a horizontal line from there to the diagonal line that represents your data. At the intersection, draw a vertical line down to the *x* axis. Read the concentration at this point on the *x* axis. This is the LD_{50} for your experiment!

Write the LD_{50} here ________________

Some Further Thoughts

LD_{50} is a sensitive way to measure the overall toxicity of a particular compound. Thankfully, the legal limits on toxic compounds in our drinking water are far, far lower than the LD_{50}. After all, you wouldn't want to drink water that kills half of the people who drink it! For example, when dealing with carcinogenic compounds, the Environmental Protection Agency requires that the amount of a toxin in the water will not raise the chance of cancer by more than 1 in 200,000 people (that's a lot less than the LD_{50}). Note that when you were working with zinc, LD_{50} was measured in milligrams per liter (mg/L). In the state of Iowa, for example, zinc must not be present in amounts more than 110 *micro*grams per liter (μg/L). A microgram is 1000 times less than a milligram. To illustrate further, and again using Iowa as an example, a few other legal limits for toxic compounds given below.

Arsenic	87.0 μg/L
Dieldrin (insecticide)	0.056 μg/L
Lead	3.2 μg/L
Nickel	52 μg/L
Nitrates and Nitrites	10.0 mg/L
Parathion (insecticide)	0.13 μg/L
Selenium	5.0 μg/L

Table 1: Zinc Concentrations and Daphnia Mortality Data

Test Concentration of zinc (mg/L)	Number Dead (your group)	Number Dead (another group)	Average Number Dead	Average % Mortality (Graph this!)
0.1				
0.5				
1.5				
4.5				

Graph: Daphnia Mortality Related to Zinc Levels

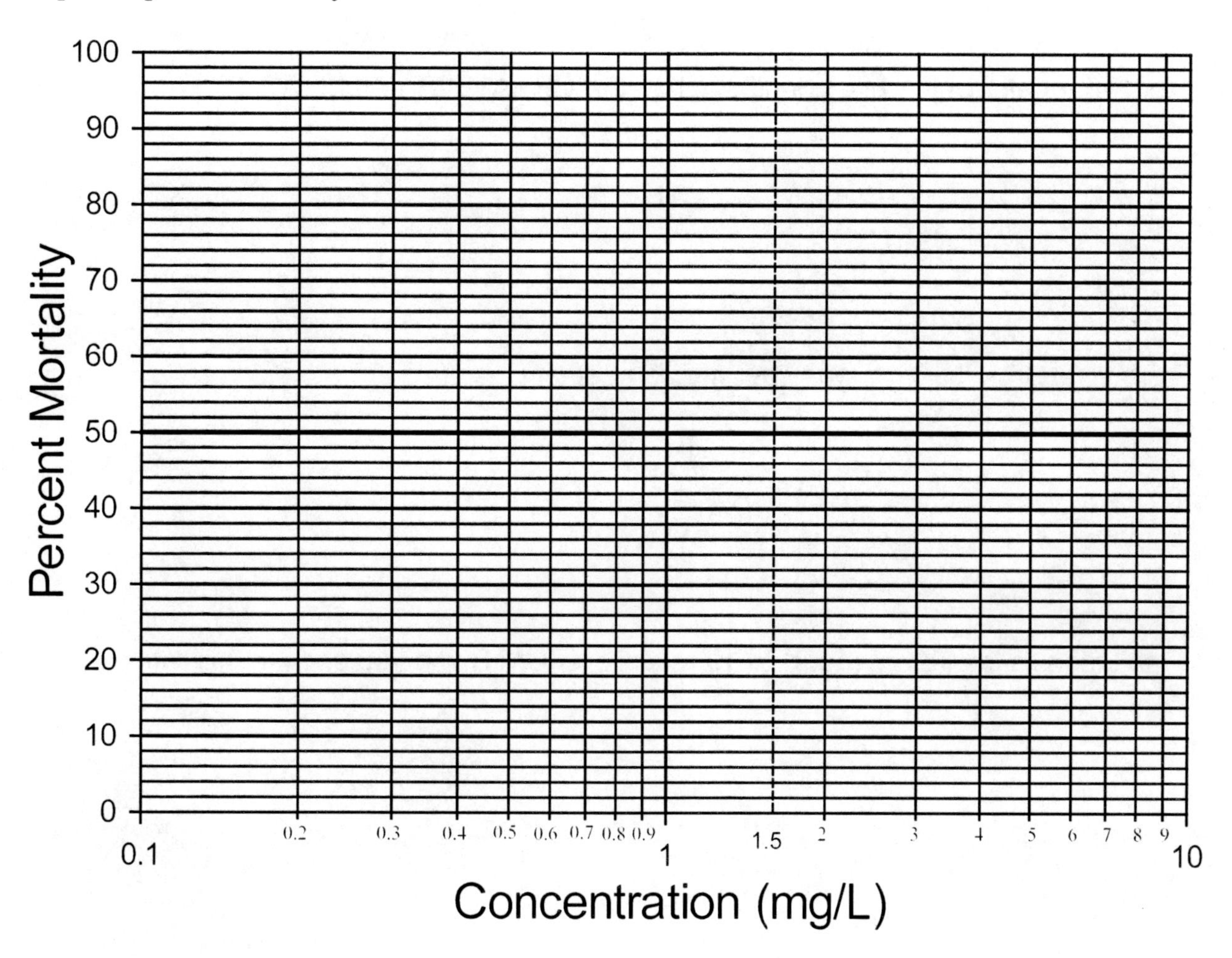

Name: ______________________________ Instructor's Name: _____________

Normal Lab Day and Time: _______________

Natural World Lab Response Sheet

Lab 10: Aquatic Toxicology

1. What was the main idea behind this week's lab?

2. From your data, which zinc conditions had the GREATEST and LOWEST % mortality?

3. From your GRAPH, what was your *Median Lethal Dose* (LD_{50})?

4. Rank today's lab from 0 (poor) to 10 (excellent). <u>WHY</u> did you choose this rating?

Questions to Consider

What does LD_{50} mean?

Where do most environmental pollutants eventually end up? Why is this?

When choosing an indicator organism for aquatic toxicity testing, should you choose one that is sensitive or insensitive to a particular pollutant? Explain your choice.

What are some characteristics of *Daphnia* that make it a good indicator organism?

Does the effect of a single pollutant on *Daphnia* give you a good idea of its effect on an entire aquatic ecosystem? What other factors should be considered?

How is the LD_{50} test useful in: (a) determining the "safe" level of a given pollutant in the environment, and (b), in comparing the toxicity of a given pollutant relative to other pollutants.

Can you think of any ways in which an LD_{50} for a species of *Daphnia* may not be a good way to determine the toxicity of a particular toxin to humans?

LAB 11: SAVING BIODIVERSITY:
AN EXAMPLE USING THREATENED COMMUNITIES*

You see before you four hypothetical *communities* of living organisms. In these communities, all objects of the same color belong to the same *taxonomic family*. Within a taxonomic family, objects of different shape represent different *species*. For example, a yellow object could be some kind of primate, and two different shaped yellow objects might be chimpanzees and orangutans. The communities you see differ in their number of individual organisms, the number of taxonomic families present, the number of species present, the number of individuals of each species, and the total weight (*biomass*) of each species.

Now, suppose that all of these communities are in immediate danger of destruction if something isn't done soon to protect them. But ... there is only enough money and political sentiment to protect one community. You are a member of a conservation team assigned the responsibility of deciding *which community will be saved.*

In real life, this conundrum is faced almost daily by conservation biologists who must make difficult decisions as to which of the many threatened communities worldwide will be conserved. How do they decide which community is most "valuable" biologically? Is it the community with the most individuals? . . . the most species? . . . the most biomass? What exactly are we trying to conserve? In this exercise, you will see that the answers to these questions concerning "biological value" are not as obvious as you might think. In fact, each of these four communities is valuable in a different way. The decision as to which one to save is, indeed, difficult. Let's approach this hypothetical problem in the way that conservation biologists might approach a real conservation issue.

Key Terms:

Community: All of the individuals of all species living together in an area.

Taxonomic family: A group of related species (like an extended family).

Species: The most closely related group of organisms.

Biomass: Mass (measured as weight) of living and dead biological material. The more biomass that is present the more productive an ecosystem is, that is, the more carbon dioxide is being converted into organisms.

Biodiversity: Consists of two parts: (1) The number of different species in a given region (*richness*). Richness sometimes can only be measured as the number of taxonomic families. (2) A measure of how equally each species is represented (*evenness*). Evenness can include the number of individuals of each species in a given area (*density*), the amount of biomass of each species, and, especially for plants, the amount of land each species occupies (*cover*).

*Developed by Mark Hafner, Louisiana State University and used with his permission.

Community assessment

Phase I: *Rapid Assessment.* In many cases, the threat of extinction is so immediate that biologists have little time to conduct an accurate species inventory of each threatened community. In these cases, biologists are forced to make a quick and subjective assessment based on available information or on a very rapid field survey or even on a reconnaissance flight in an airplane or helicopter. Remember, any delay at this stage could result in loss of all the communities.

To simulate a Rapid Assessment, simply examine the four communities visually for a minute or two (approximately 30 seconds per community), and then make a quick decision as to which community you will save. Do not touch the objects or discuss the communities with your classmates. (Don't worry, there's no single "correct" answer.) Before moving to Phase II, record your rapid assessment choice and reason for that choice.

Phase I (Rapid Assessment) Choice: ______________________

Reason:

Phase II: *Community Inventories.* Assuming that time is available, biologists will conduct an inventory of the number of individual organisms, species, genera, etc. living in each community. These base-line data will be used to help determine which community to save.

You are now a member of a biological inventory team assigned to one of four communities. You must determine the number of species, the number of individuals per species, and each species' total biomass. You can now touch the objects. Fill in the blanks in the following list. Then complete the two tables on the next page. Use the electronic scale to weigh biomass in grams. Measure to the nearest 0.1 gram.

Biological inventory of community #_______

1. Community density (total number of individuals in the community): _________
2. Species richness (total number of species in the community): _________
3. Taxonomic richness (total number of families in the community): _________

As a team, fill out the tables on the next page to calculate Shannon's Index of Diversity based on density (number of individuals) and biomass for your community. Shannon's Index gives you a single number that takes into account both components of diversity: richness and evenness.

Table 1: Shannon's Index based on density: Use a calculator to find the natural log (*ln*) of p_i, where p_i, the proportion, is the number of individuals of species "I" divided by the total number of individuals of <u>all</u> species. The negative sum of $p_i(\ln(p_i))$ is the Shannon Diversity Index. Round your numbers to two decimal places.

Species description	Number of Individuals	p_i	$\ln(p_i)$	$p_i(\ln(p_i))$
	Total=			-Sum=

Table 2: Shannon's Index based on biomass: As above, use a calculator to find the natural log (*ln*) of p_i, where p_i is the biomass of species "I" divided by the total biomass of <u>all</u> species.

Species description	Biomass (g)	p_i	$\ln(p_i)$	$p_i(\ln(p_i))$
	Total=			-Sum=

Phase III: *Community Comparisons*. When the inventories are complete we will compare the four communities. Notice that each of the factors listed below measure diversity in a different way. The Shannon's Diversity Index provides a single number that represents both species richness and evenness. It may assist you in making a more objective comparison between communities. The higher the Index number, the greater the diversity.

As a class, fill in the appropriate data on the table below. (Your instructor will have a master data sheet on the overhead projector). As a group, rank the four communities (highest to lowest) for each of these factors to determine which community is the most diverse based on each particular factor. Each group will come to its own consensus as to which community should be saved. In the space on the next page, each student will write three reasons for saving that community.

	Rank				Community #1	#2	#3	#4
Community Density (number of individuals/community):					____	____	____	____
Rank by density	____ high	____	____	____ low				
Species Richness (number of species/community):					____	____	____	____
Rank by species richness	____	____	____	____				
Taxonomic Richness (total number of families):					____	____	____	____
Rank by taxonomic richness	____	____	____	____				
Total Community Biomass:					____	____	____	____
Rank by biomass	____	____	____	____				
Shannon's Diversity Index (density):					____	____	____	____
Rank by Index	____	____	____	____				
Shannon's Diversity Index (biomass):					____	____	____	____
Rank by Index	____	____	____	____				

Phase III Choice: ____________________

Reasons:

1.

2.

3.

Think about your "second" choice. Why did you end up not picking that community? Give a few of your reasons.

Phase IV: *The Decision.* This is the difficult part. As a class, we will discuss how a team of biologists might assess the relative "biological value" of these four hypothetical communities.

Can we reach a class consensus on which community to save based solely on the information provided? Record your class consensus, if any.

Class Consensus: ________________

Phase V: *More Information.* To add more realism to the exercise, on page 6 are the results of several "research projects" conducted on the communities you have been studying. These results of these projects are presented as "News Flashes." You, the conservation biologist, must now consider these new data in this difficult, yet critically important decision.

How does this information influence *your* decision? In the space below, write your groups' final decision for which community should be saved. State whether or not the news flash played a role in your final decision. Why did/didn't the news flash make a difference?

The Final Decision: ________________________

Did the "News Flash" influence your decision? Why or why not?

NEWS FLASH!

New research coming from the field has just revealed that:

Community 1. The four yellow balls represent the last two pairs of Gorillas living in the wild. Note: The Gorilla is the largest species of primate that has ever lived on earth.

Community 2. The long, gray objects represent mongoose-like mammals that are important ceremonial species to local tribesmen (they represent mythical gods). These mammals also keep the local rat and snake populations under control.

Community 3. The blue cubes represent a species of rodent that hosts a very rare species of fungus. This fungus has been used for medicinal purposes for centuries by local Indians and has the potential for combating certain kinds of cancer in humans.

Community 4. The six largest objects represent a species of exotic bat with a majestic 5-foot wingspan. This noble species is found elsewhere but it is always very rare. These bats mate for life, and each pair has only one offspring every two years.

Name: ______________________________ Instructor's Name: ____________

Normal Lab Day and Time: ______________

Natural World Lab Response Sheet

Lab 11: Assessing Biodiversity

1. What was the main idea behind today's lab?

2. What are the 2 sub-divisions of biodiversity? How do they differ from each other?

3. Below are data on Shannon's Diversity Index (density) for 4 communities. Using this value only, circle which community is most diverse.

Community 1 = 0.75　　　Community 2 = 1.20

Community 3 = 0.93　　　Community 4 = 1.43

4. Two ecological communities, A and B, have the same number of species, but community B has a lower Shannon's Diversity Index. This indicates that community B has a lower __________.

A. evenness　　B. taxonomic diversity　　C. biomass　　D. density

5. Rank today's lab from 0 (poor) to 10 (excellent). WHY did you choose this rating?

LAB 12: PUZZLING THE PAST

You've most likely seen a "tree" depicting the hypothesized evolutionary history of some group of organisms. Perhaps it looked like this 19th Century example from the biologist Ernst Haeckel (below left). Or this one (below right), the *only* illustration from Charles Darwin's book *On the Origin of Species*:

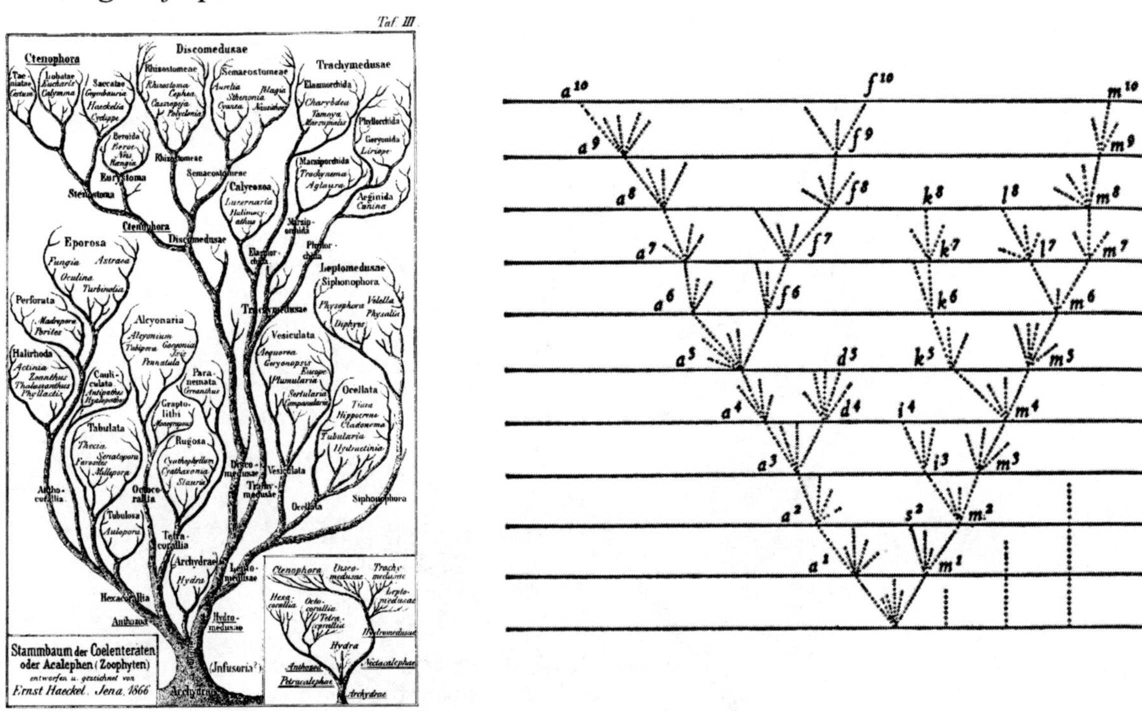

The earliest known evolutionary tree (below left) was made by Jean Baptiste Pierre Antoine de Monet, Chevalier de Lamarck (usually shortened, thankfully, to Lamarck) published in 1809, long before Darwin came up with the *mechanism* of evolution called Natural Selection.

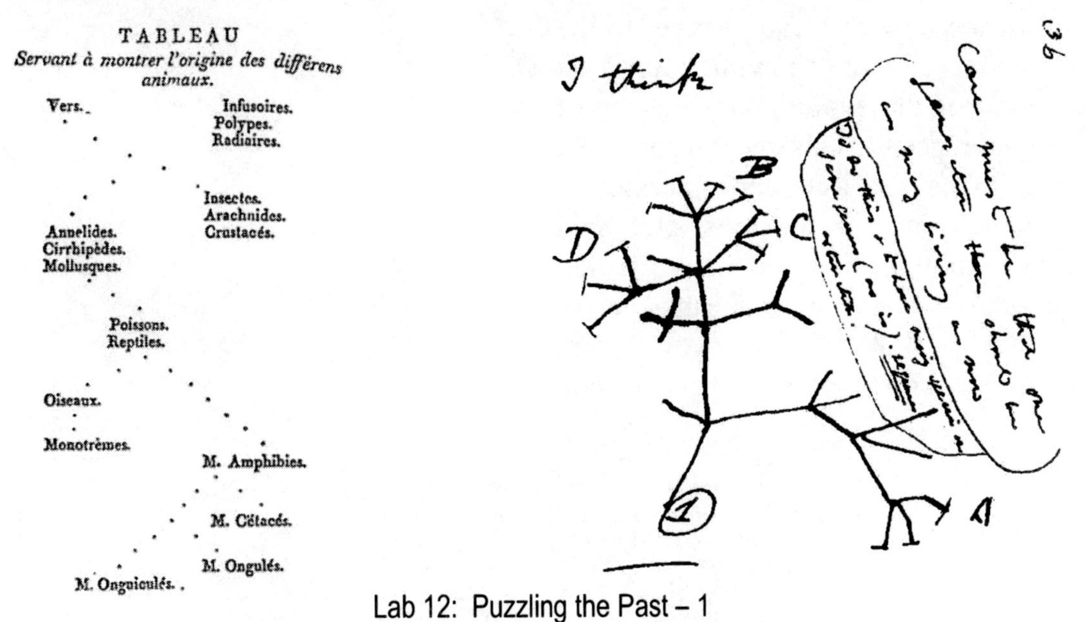

Darwin's first known tree (above right) is found in one of his notebooks. It was drawn early in his career while he was still working out his theory of natural selection. Although others thought, or were beginning to think, that evolution occurs, Darwin and Alfred Russell Wallace were the first to come up with a plausible mechanism of evolution. Note where Darwin writes "I think".

Although we now draw evolutionary trees much differently than in these examples, modern trees are still meant to convey the same information – namely evolutionary history. Modern trees typically look like this one. Its various parts have been labeled.

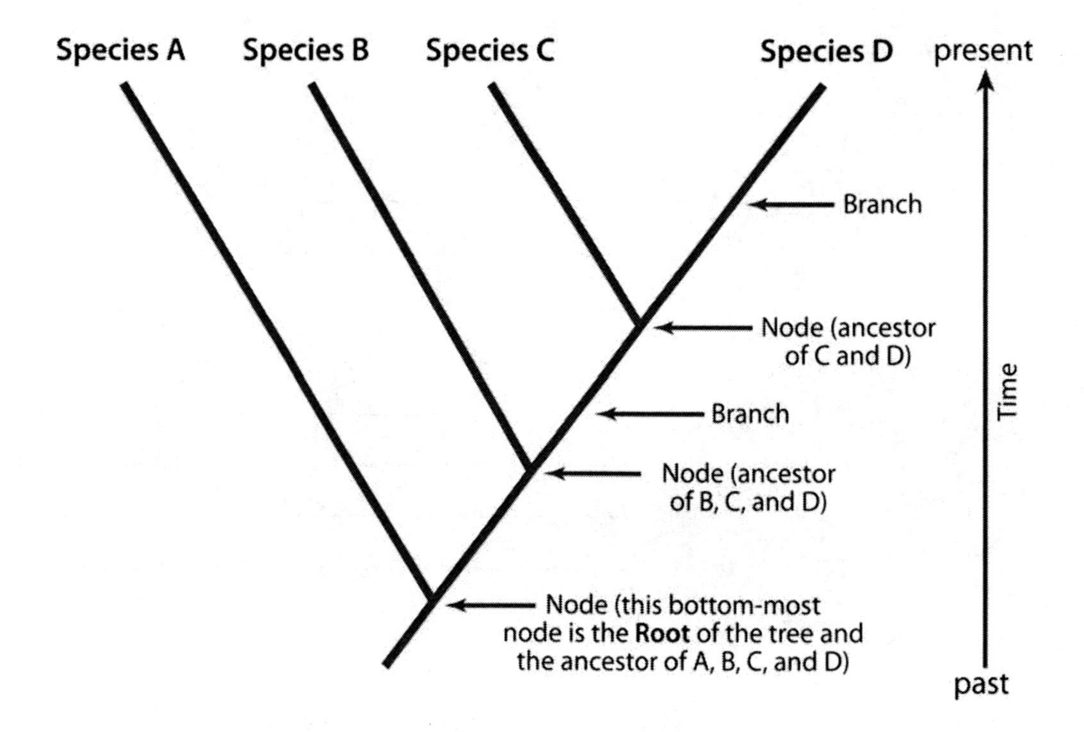

To illustrate how to "read" a tree, to the right is given a tree showing our best understanding of the evolutionary history of several primate species. This tree indicates that chimps + bonobos share a common ancestor, as do humans + chimps + bonobos, etc. Said another way, chimps and bonobos are *sisters* and humans are sister to the lineage made up of chimps + bonobos. We did *not* evolve from chimps and bonobos, we share a common ancestor.

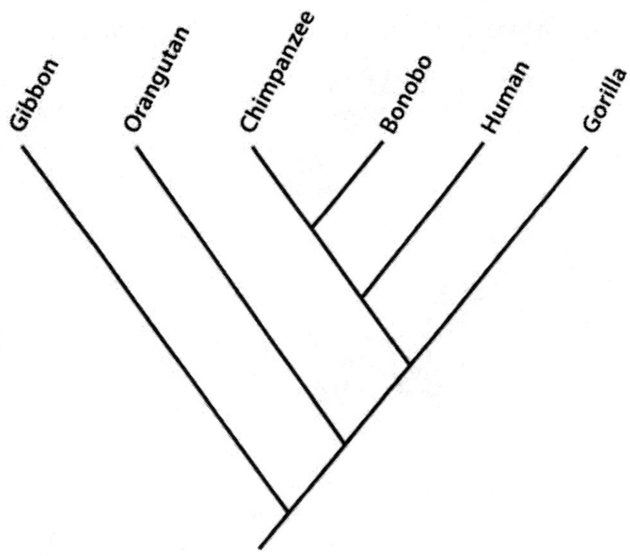

Creating a Phylogenetic Tree

Often we refer to an evolutionary tree as a *phylogenetic tree*, that is, a drawing that depicts *phylogeny* – evolutionary history. Scientist regularly construct phylogenetic trees of the groups of organisms that they study. How these trees are constructed, however, probably looks mysterious to the lay person. In this investigation you will be learning a common method that biologists use to reconstruct evolutionary history and how to convey this information using a phylogenetic tree.

Consider the possible ways that just three species (A, B, and C) can be related. All possibilities are given below. There are no other possibilities. Note that the nodes of a tree can swivel. For example, in the third tree the order of species shown as C, A, B says the same thing as C, B, A).

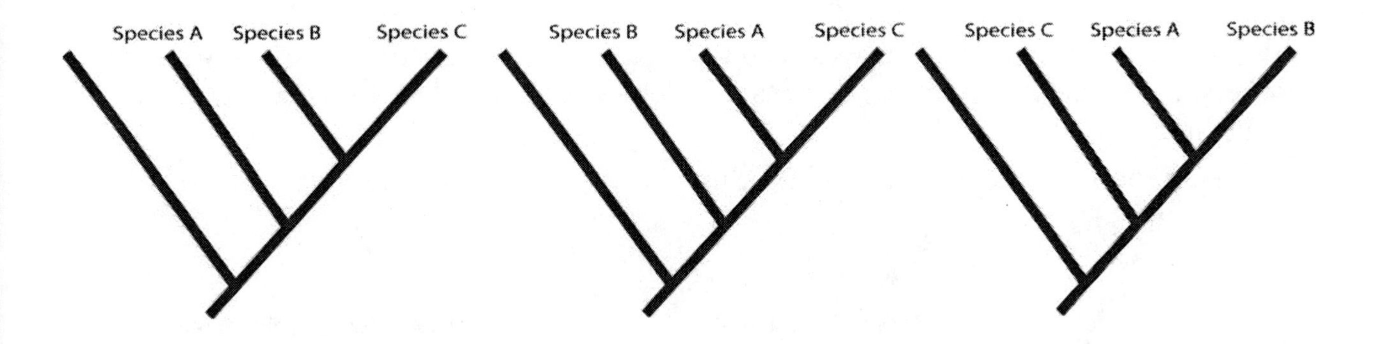

Although three species can be related in only one of three possible ways, for four species there are 15 possible ways; for 5 species there are 105 possible trees; for 10 there are 34,459,425 trees; for 20 there are 820 billion billion billion possible trees! Yikes. How do you choose the best one? Obviously, for more than a few species you need a computer to get the job done, and even then shortcuts must be taken and not every tree can be examined.

But back to our three species problem. We need to collect some data. Presume that we are working with three plant species and come up with the following data matrix – a very small matrix indeed. Note that we're also gathering data for at least one *outgroup*, a species or group of species that with out a doubt lie outside of the group you're studying. The outgroup is used to *polarize* the characters, that is, to determine the direction of evolution – the *ancestral* (older) versus the *derived* (newer) state of a particular character (trait). We assume that the characters in the outgroup posses the ancestral condition.

Species	Flower Color	Leaves per Node	Seeds per Fruit	Chromosomes	Number of Petals
Outgroup	white	1	10	10	3
A	white	2	10	10	4
B	red	2	10	20	4
C	red	2	15	20	3

Now, how to pick the best of our three possible trees? As in all scientific endeavors, we are going to apply *Occam's Razor*, which says that the simplest explanation – the one that requires the fewest hypotheses – is the best one. Said another way, we want the most *parsimonious* (simplest) explanation. Keep in mind that *any* evolutionary change is highly unlikely, so it is *extremely* improbable that any particular evolutionary change will independently happen more than once in the history of a group of organisms. Let's take Tree #3 above and map onto it each of the changes implied by reference to the outgroup. Changes on the tree are called ***steps***, these can be single mutations, changes in behavior, losses or gains of DNA, new morphologies, change in habitat, etc.

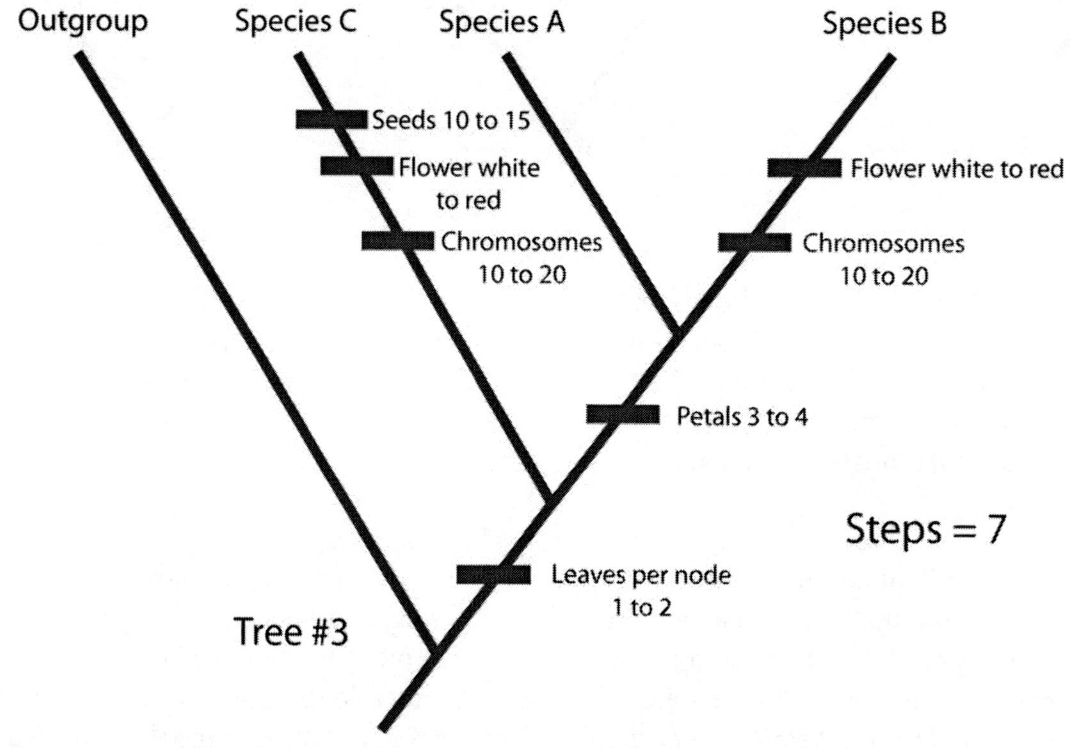

Note that there are 7 hypothesized evolutionary changes (steps) and that two of them had to happen twice. (Your instructor will help you out with this if some confusion remains.) Observe that the "Leaves per node from 1 to 2" character unites species A, B, and C into a single lineage (we say *clade*); likewise the "Petals from 3 to 4" character unites species A and B into a clade. These shared, derived characters are called *synapomorphies*.

Below are the remaining trees (#1 and # 2). Map the characters onto these trees as in the example above. **Note that if you cannot "evolve" character only once on the tree, then you must evolve it twice** (or more) ***or*** **evolve it once and then have it revert back to a previous condition.**

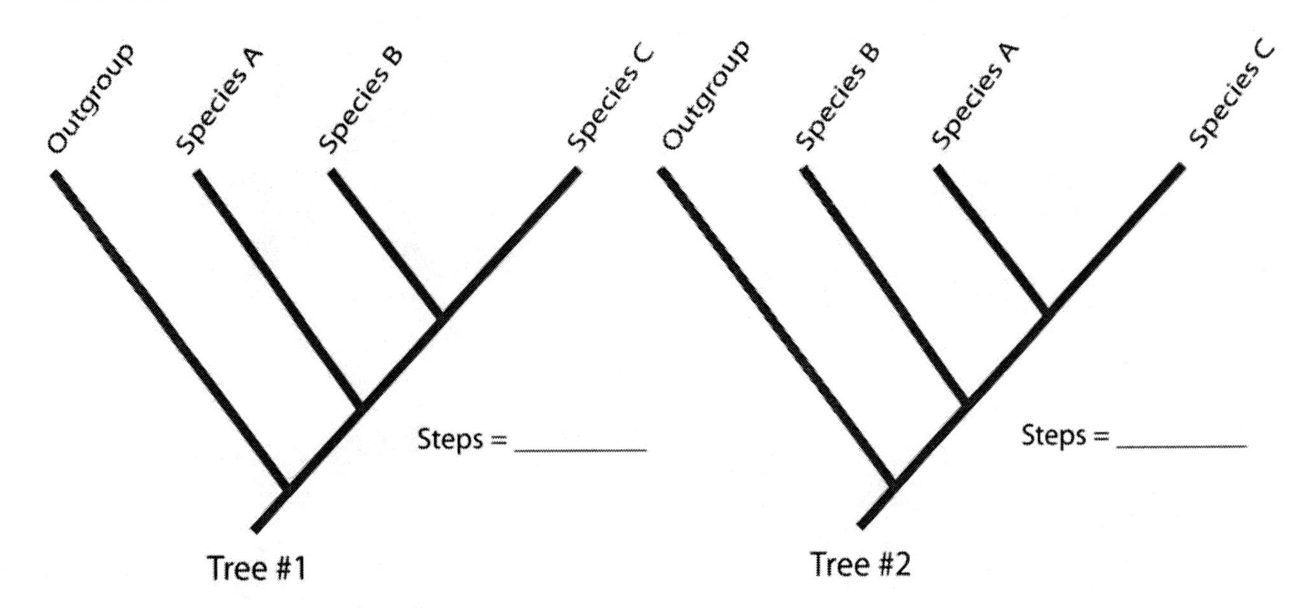

Which of the three trees had the fewest steps? Write the tree number here: ________. This tree requires the fewest number of evolutionary hypotheses and is, therefore, the *most parsimonious tree* (MPT). (Again, if you're having trouble mapping characters on the trees your lab instructor will help you out.) The MPT is the best estimate of the evolution of the three species of interest under the criterion of parsimony.

Activity: Reconstructing Evolutionary History

In the following Investigation you must do your best to estimate the evolutionary history of twelve hypothetical "species." With 12 "species" to work with, there are clearly many (many!) millions of possible phylogenetic trees! So this is not a simple exercise. Look for shared derived characters (*synapomorphies*) that will unite two or more species within a single branch of a phylogeny. There is probably no single best tree because your group will likely define a particular character differently than does another group of students. When this lab was developed, however, its creators had a single character definition in mind. Based on *their* definition of a character, there is but one best phylogeny. See if you can find it.

Procedure:

1. ☐ Each group of students has been given plastic models of 12 hypothetical "species," including an outgroup, as illustrated below. Use the illustration to take notes if you'd like (tear the page out if it helps). Your groups must first decide what the particular characters (traits) of the organisms are.

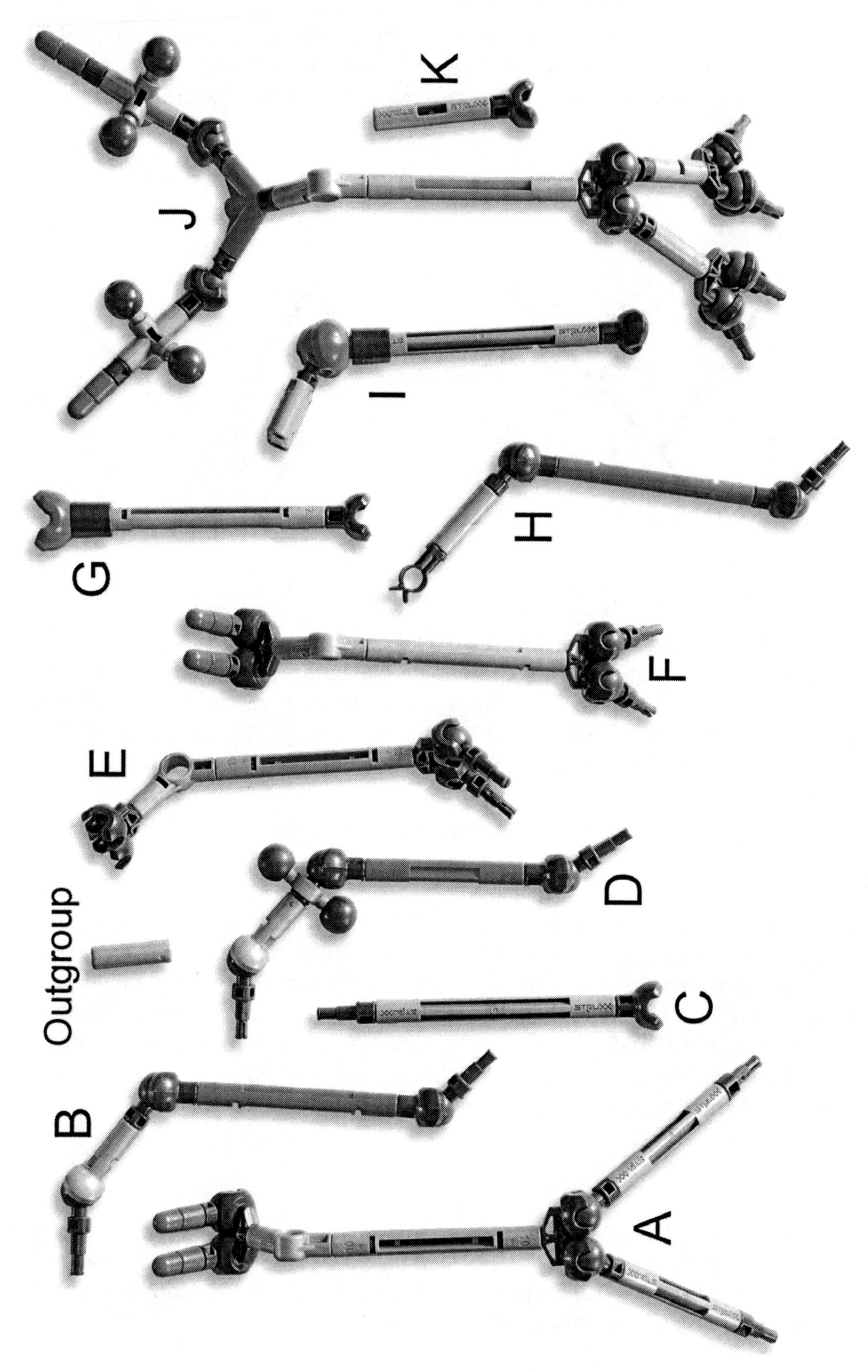
Outgroup
A
B
C
D
E
F
G
H
I
J
K

2. ☐ Your job is to reconstruct the evolutionary history of these species based on the reasoning presented above: the fewer "steps" the better. Use the blank pages below to take notes and to **draw your estimate** of the phylogenetic tree. Remember, we cannot see ancestors. Use a pencil – you *will* need to erase!

Keep in mind that we can never (or at least very very rarely) identify actual "ancestors" (the nodes); we can only study living organisms and fossils that are probably "cousins" rather than direct ancestors. The chance of an "ancestor" being fossilized is close to zero. That's a fancy way of saying that observable organisms are at the tips of branches, not at the nodes.

Note: This exercise is not as simple as it seems! You'll have to work together to come up with a plausible solution.

Hint #1: the first few "real" branches of the phylogeny are given below.

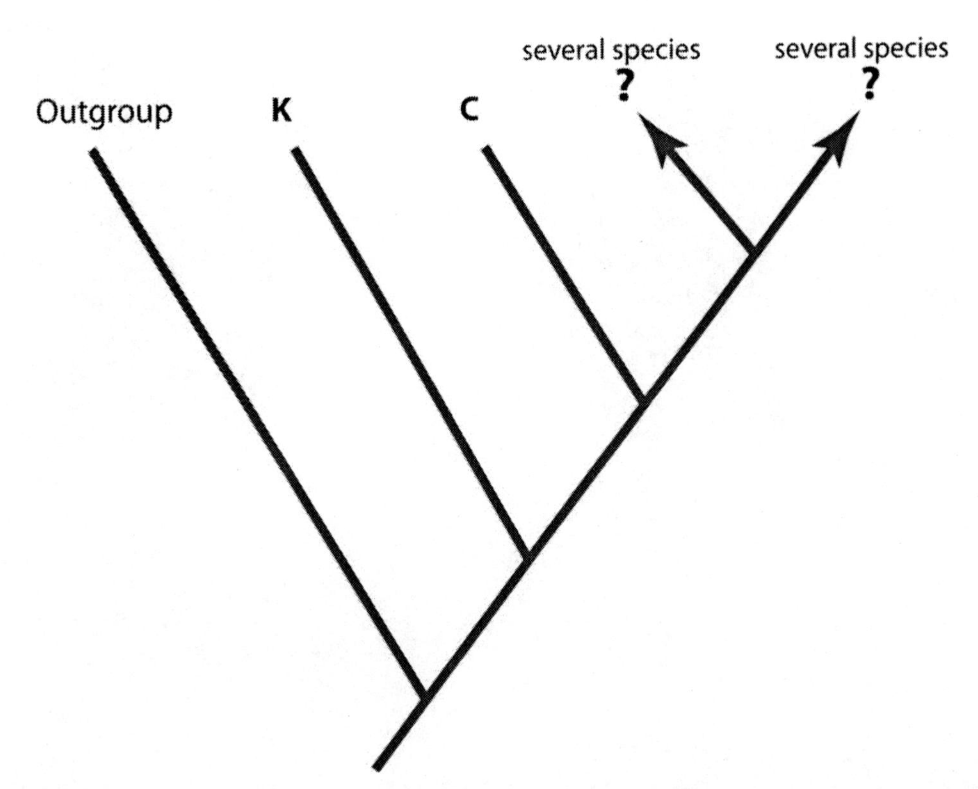

Remember, the outgroup possesses the ancestral condition of a trait. In this case the outgroup "species" consists of only a short "tube" (a body?). K, the first "species" to branch off, has the new features of a "cup" and a longer "tube." In fact, all species on the rest of the tree share the "cup" and the longer "tube" – these are synapomorphies. The next species (C) to branch off gained a "snout." All remaining species also have a snout so this character is a new synapomorphy. Keep looking for synapomorphies to build the rest of the tree.

Hint #2: The next two synapomorphies to consider are an orange (versus a grey) "tube" and the presence (versus the absence) of a black "ring" on the body.

Notes and Tree (then go to page 10):

3. ☐ As a class, discuss your various solutions to solving the best plausible phylogeny of the species you examined. Be sure to defend your choice of what was and what wasn't a "character."

Questions to Consider

Why was it difficult to come up with a single, best phylogeny of these species?

Assuming that the "species" are real organisms, what kind of information would help you decide what a particular character really was?

Why do scientists rely on parsimony to focus their results?

Could the concept of parsimony lead to an incorrect hypothesis? Explain.

What did you learn about how scientists approach the problem of phylogeny reconstruction?